Math Resources
from Recyclables

by
Bonnie Mertzlufft • Brenda Morton • Virginia Woolf

illustrated by
Susan Pinkerton

Publisher: *Roberta Suid*
Editor: *Carol Whiteley*
Design and Production: *Susan Pinkerton*
Cover design: *Terrence Meagher*
Consultants: *Catherine Dilts* and *Lillian Lieberman*

For a complete catalog please write to the address below.
Monday Morning Books
P.O. Box 1680, Palo Alto, CA 94302

ISBN 1-878279-70-X

Printed in the United States of America

987654321

CONTENTS

INTRODUCTION

Math Resources from Recyclables brings the concepts of recycling and educating together. This book focuses on reusing everyday materials in activities and projects that further many mathematic skills: addition, subtraction, estimating, time-telling, and more. By making learning tools from throwaways, you will be conserving your own time and energy while helping to save the environment. Most of the projects in this book may be "recycled" in your own classroom in the years ahead.

As you flip through the pages of *Math Resources from Recyclables,* you will notice the easy-to-follow manner in which the projects are presented. Each activity includes a list of materials, detailed directions for constructing the item, and suggested methods for use with students in your classroom. Reproducible patterns, worksheets, samples, and illustrations are also provided throughout the book.

Math Resources from Recyclables is divided into thirteen chapters. Each chapter is devoted to the use of a different recyclable material, including newspapers, small milk cartons, egg cartons, old calendars, small boxes, and more. The activities in each chapter cover four levels of skill development, geared toward the primary grades (K-3). However, you should not feel limited by the grade levels suggested on the activity pages. Many of the projects can easily be adapted to fit the needs of other grade levels, other subjects across the curriculum, or children with special needs.

As a concerned educator, you have probably looked for breaks from the daily paper/pencil classroom routine. These activities will help involve students in the learning process by encouraging them to manipulate activity pieces, cut, paste, work creatively, and more. The students will also be able to work away from their desks on many of the projects. Use the floor, the reading table, or the bulletin board as alternate working surfaces to make the school day more fun and interesting for your students.

Your rewards from using the activities in *Math Resources from Recyclables* will be great! Not only will you help the environment by recycling and save your own time by creating long-lasting learning equipment, most importantly you'll watch your students achieve in mathematics.

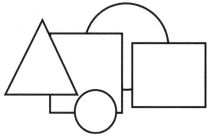

PARENT LETTER

Note to Teacher:
Duplicate this letter and place a check mark beside the items you need. Send the letter home with your students.

Dear Parents,
     We are making tools from recyclables to use in our classroom. If you have any of the following items, please send them to school as soon as possible.

_____ milk jug caps (all colors)
_____ Styrofoam meat trays (all sizes and colors)
_____ old magazines
_____ large paper grocery bags
_____ empty boxes (all shapes and sizes)
_____ old calendars
_____ newspapers
_____ used wrapping paper
_____ Popsicle sticks
_____ cardboard egg cartons
_____ used greeting cards

     If you are available to help assemble resources or work with children in the classroom, please mark the appropriate spaces below.

_____ I can help at school (_____ mornings/_____evenings).
_____ I can help at home.
The best day for me is _____.

Thank you!

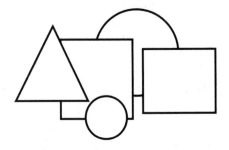

PARENT THANK YOU NOTE

Note to Teacher:

Duplicate this thank you card onto colored paper. Send it home with children who have brought supply materials for the recycling projects. You might also send it to parents or guardians who have donated their time to help.

Recycling Counts!
I was glad I could count
on you. Thank you.

_____
teacher signature

CHAPTER ONE

MILK JUG CAPS

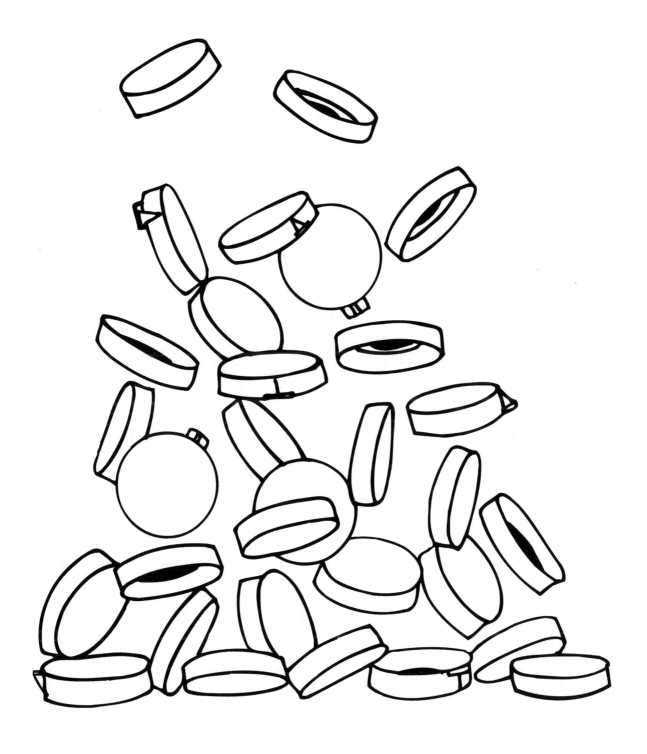

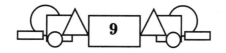

NUMBER CONCEPTS

KINDERGARTEN

Objective: To reinforce number concepts 1-5 by matching objects to symbols.

Materials: milk jug caps in sets of 25, numeral cards (p. 11), 15 small buttons, 15 popcorn kernels, 15 beans, 15 screws, 15 pebbles, plastic bag, tagboard, scissors, tape, glue

Construction:
1. Duplicate, mount on tagboard, and laminate the numeral cards.
2. Cut the circles out and tape them to five of the caps.
3. Glue one of each item (buttons, popcorn kernels, beans, etc.) on five different lids.
4. Continue gluing two, three, four, and five of each item on the remaining lids. You will end up with lids with one button, two buttons, three buttons, four buttons, and five buttons. Repeat with the other materials.
5. Store sets in a plastic bag.

Directions for Student:
1. Put the five numbered lids in a row.
2. Place the lids with one item beside the number 1 lid.
3. Place the lids with two items beside the number 2 lid, the threes with 3, and so on.

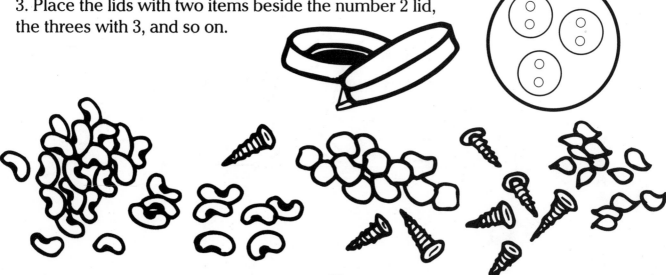

NUMERAL CARDS

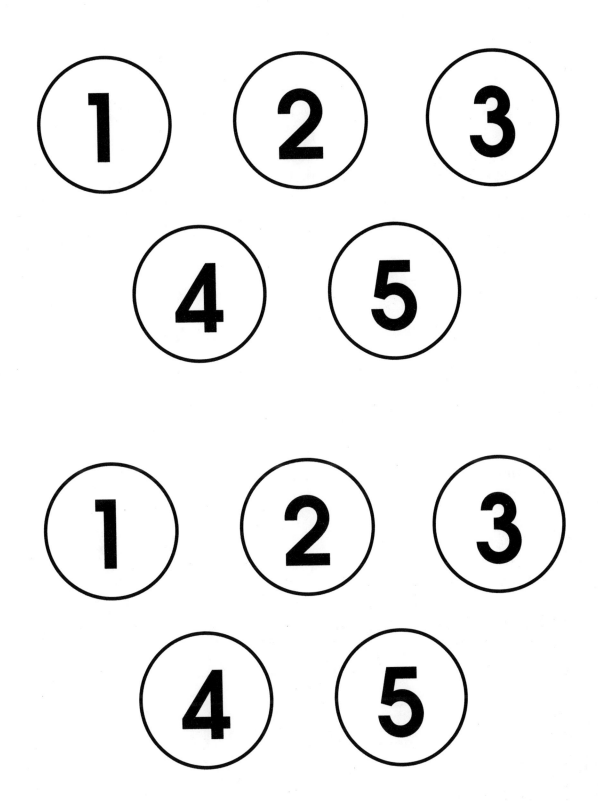

NUMBER SEQUENCING

FIRST GRADE

Objective: To reinforce counting by fives.

Materials: milk jug caps in sets of 10, 10 circles cut from tagboard scraps, marker, tape, scissors, plastic bag

Construction:
1. Write one numeral on each circle, counting by fives from 5 to 50.
2. Laminate and cut out the circles.
3. Tape the circles to the milk jug caps.
4. Store the caps in a plastic bag.

Directions for Student:
1. Place the milk jug caps onto a flat desk or on the floor.
2. Arrange the caps in order by fives. Start with the number 5 and end with the number 50.

Option: You can also duplicate this activity for 2's and 10's practice.

PLACE VALUE

SECOND GRADE

Objective: To reinforce the concept of place value.

Materials: milk jug caps, 30 circles the size of milk jug caps cut from scraps of tagboard, scissors, markers, tape, plastic bag, student score card (p. 14)

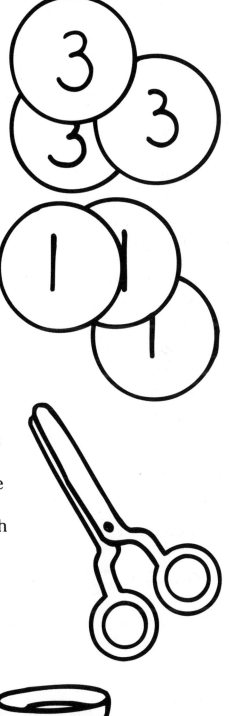

Construction:
1. Write each number 0 through 9 on three circles. (You will have three circles with 0, three with 1, three with 2, and so on.)
2. Laminate the number circles, cut out, and tape to the milk jug caps.
3. Duplicate the student score card.
4. Store all pieces in a plastic bag.

Directions for Two Students:
1. Place the milk jug caps number-side down on a flat surface.
2. Take turns choosing one of the milk caps.
3. Write the first number that you choose in the ones column of the score card, the second number in the tens column, and the third number in the hundreds column.
4. After filling out your first three-digit number, see whose number is highest. Circle the higher number.
5. When you have finished filling out the sheet, see which of the circled numbers is the highest. Which number is the lowest?

STUDENT SCORE CARD

Student 1. _____

	Hundreds	Tens	Ones
1.	___	___	___
2.	___	___	___
3.	___	___	___
4.	___	___	___
5.	___	___	___

Student 2. _____

	Hundreds	Tens	Ones
1.	___	___	___
2.	___	___	___
3.	___	___	___
4.	___	___	___
5.	___	___	___

HOW MANY MINUTES?

THIRD GRADE

Objective: To reinforce time-telling skills.

Materials: 20 milk jug caps, clock patterns (p. 16), scissors, tagboard, glue, plastic bag, paper, pencils

Construction:
1. Duplicate the clock patterns onto tagboard, laminate, and cut out.
2. Glue one clock pattern to each milk jug cap.
3. Store all caps in a plastic bag.

Directions for Student:
1. Number your paper from 1 to 10.
2. Pull out two clocks at a time from the plastic bag.
3. Write down the two times on your paper and then write the amount of time that has gone by from the first clock to the second. For example, if the time on the first clock is 12:00, and the time on the second is 3:30, then three and a half hours have gone by.

Note: Tell the children that all times are either a.m. or p.m. for less confusion.

1. 12:00 - 3:30 = 3 and a half hours
2. _____
3. _____
4. _____
5. _____
6. _____
7. _____
8. _____
9. _____
10. _____

three and a half hours have gone by

CLOCK PATTERNS

CHAPTER TWO

STYROFOAM MEAT TRAYS

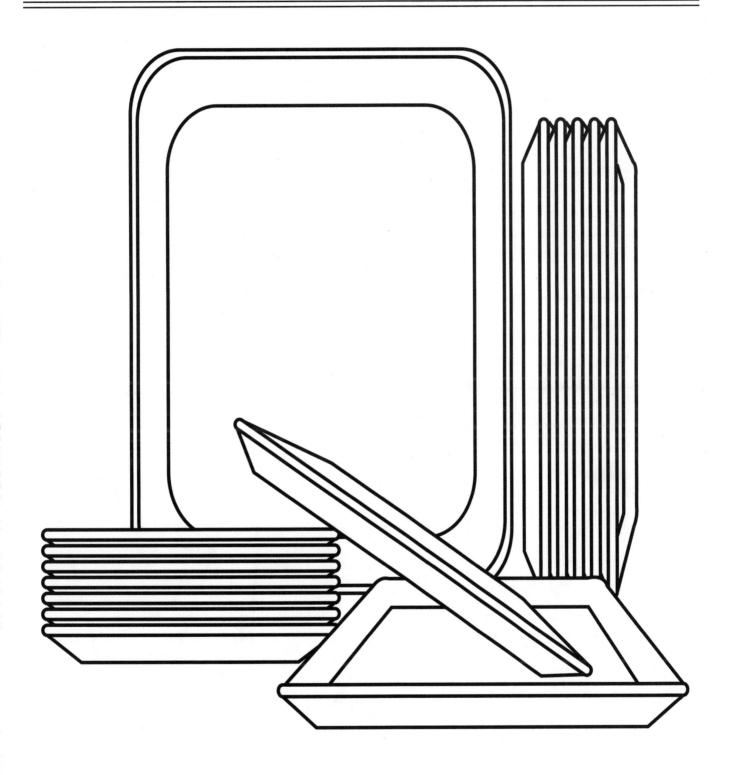

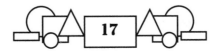

SHAPE STENCILS

KINDERGARTEN

Objective: To reinforce shape designs and names.

Materials: meat trays in sets of 4, shape patterns (p. 19), scissors, pencils, paper

Construction:
1. Cut out the center of each meat tray in the shape of a circle, triangle, rectangle, or square.
2. Help the children write the name of the shapes on their papers if they require assistance.

Directions for Student:
1. Use the stencils to draw the shape on your paper.
2. Write the name of each shape next to it.
3. Turn the shape into a simple picture by drawing. For example, a circle can be a happy face, sun, or flower. A square can be a house or a present. A triangle can be a hat, volcano, or piece of cheese. And a rectangle can be a truck, wagon, or bar of chocolate!

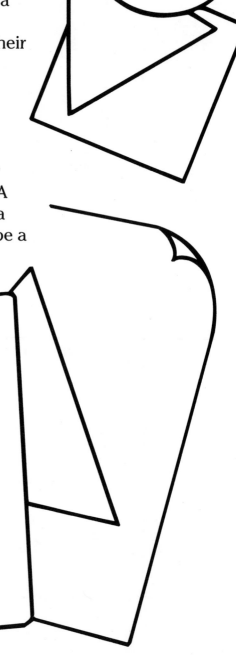

SHAPE PATTERNS

MATH MANIPULATIVES

FIRST GRADE

Objective: To reinforce addition skills.

Materials: divided meat trays (1 per student), small strips of construction paper, marker, scissors, tape, manipulatives (beans, popcorn, beads, buttons, or pebbles), nut cups or small paper cups

Construction:
1. Write an addition fact on each strip of construction paper, but do not give the answer.
2. Laminate the strips and cut them out.
3. Tape one card into the long section of each tray.
4. Place the counting items in the cups and place them in the middle of the table where students will be doing this activity.

Directions for Student:
1. Look at the addition problem on the tray.
2. Place the correct number of counting items in the first section of the meat tray to show the first addend.
3. Count out the second addend with the items and place them in the second half of the meat tray.
4. Add both sections together to find out the answer to the math fact.

Note: Math fact cards may be numbered, and the child can then record the numbers and answers on a piece of paper. Children can trade trays until they complete a certain number you designate, with answer cards provided.

MISSING ADDEND

SECOND GRADE

Objective: To reinforce addition and subtraction skills.

Materials: 4 meat trays, craft knife, permanent marker, scissors, paper, pencils

Construction:
1. Use the permanent marker to write an incomplete addition problem on each meat tray—the problems should go across, for example,

$$3 + \underline{\quad} = 6$$

2. Use the craft knife to cut out a box in the space of the missing addend.
3. Number the meat trays from 1 to 4.
4. Cut paper for the students to fit the meat trays.
Note: You may put the answer on the back of each tray for self-checking.

Directions for Student:
1. Number your first paper 1.
2. Place the paper underneath the first meat tray and fill in the missing number that will make the problem work.
3. Put a number 2 on another piece of paper. Place the paper under the second meat tray and fill in the missing number. Do the same thing for trays 3 and 4.
4. Check your papers against the teacher's answer key.

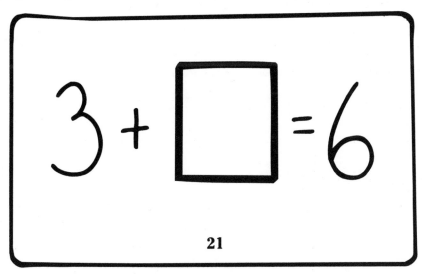

METRIC MEASURING

THIRD GRADE

Objective: To practice measuring in centimeters.

Materials: 4 large meat trays (all the same size), ink pen, various lengths of yarn in a variety of colors, large needle, centimeter ruler, scissors, paper, pencils, small box

Construction:
1. Use an ink pen to write the numerals 1 to 5 down the side of the flat surface of one meat tray. Write 6 to 10 on the next one, 11 to 15 on the next, and 16 to 20 on the last one. Note: Be sure not to press so hard that you poke holes in the meat tray.
2. Use the needle to make a hole next to each number.
3. Use a centimeter ruler to mark an exact length from each hole to another spot along the length of the tray.
4. Punch another hole at each of the marked spots.
5. Thread a different-colored length of yarn between each pair of holes and tie the yarn in the back. (The students will measure the yarn lengths.)
6. Store the meat trays in the box.

Directions for Student:
1. Number a paper from 1 to 20.
2. Use a centimeter ruler to measure the length of each piece of yarn to the nearest centimeter.
3. Write down the answer beside the correct number on your paper.
4. Give your completed paper to the teacher for checking.

CHAPTER THREE

SMALL MILK CARTONS

MARBLE TRAINS

KINDERGARTEN

Objective: To reinforce counting and numeral-recognition skills.

Materials: 10 milk cartons with tops cut off, 55 marbles, yarn, hole punch, construction paper, scissors, glue, zip-lock plastic bag

Construction:
1. Punch two holes in each carton, one on the front and one on the back.
2. Attach the cartons with lengths of yarn threaded through the holes.
3. Cover the milk cartons with construction paper.
4. On the side of each train car, write one number, from 1 to 10.
5. Store the marbles in the plastic bag.
6. After the children have placed the same number of marbles in each car as the number on the car, let them play with the train.

Directions for Student:
1. Place the same number of marbles into each train car as the number on the side of the car.
2. Show the teacher the train when you have finished placing the marbles.

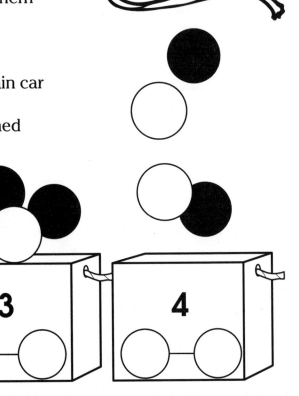

THE FIVE AND DIME

FIRST GRADE

Objective: To reinforce money concepts.

Materials: 12-15 empty milk cartons, marker, coin patterns (p. 26), scissors, tagboard, plastic bag, construction paper, glue

Construction:
1. Cover the milk cartons with construction paper.
2. Label each milk carton with an amount of money in five cent increments (5, 10, 15, 20, 25, and so on).
3. Duplicate the coin patterns onto tagboard, laminate, and cut them out.
4. Store the "coins" in a plastic bag.

Directions for Student:
1. Look at the money amount on each milk carton.
2. Put the matching amount of coins in each carton, using dimes and nickels.
3. Show the teacher the cartons before putting the coins back into the plastic bag for the next student's use.

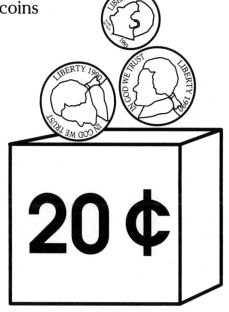

COIN PATTERNS

MILK CARTON CREATURES

SECOND GRADE

Objective: To practice math facts.

Materials: one small milk carton (clean and dry) per child, construction paper scraps, glue, markers, pipe cleaners, scissors, thin paper strips, pen, scissors or sharp knife

Construction:
1. Using a sharp knife or large scissors, make a slit in the front panel of each milk carton for the math facts strip.
2. Have each child open up his or her milk carton to form a box, and turn the box so that the opening is at the bottom.
3. Encourage the children to decorate their cartons to look like robots, aliens, animals, or whatever they desire. They can glue pipe cleaners to the sides for arms.
4. Make math fact strips based on the level of your students. Follow the example on this page. Write the answers on the back for self-checking.
5. Give each child a math fact strip to insert through the "mouth" of his or her creature, with the end rolled up inside the carton.

Directions for Student:
1. Pull out the paper strip so that one fact is shown.
2. Say the fact answer and then check to see if you are correct. Continue to anwer every problem.
3. Make a list of any answers you miss to practice.
4. Take your creature home for additional math practice.

MAKING CHANGE

THIRD GRADE

Objective: To reinforce addition and subtraction skills when dealing with money.

Materials: 5 small milk cartons, 5 objects that will fit inside the milk cartons, price tag patterns (p. 29), money patterns and coin patterns (p. 26 and p. 30), marker, colored Contact paper, scissors, plastic bags

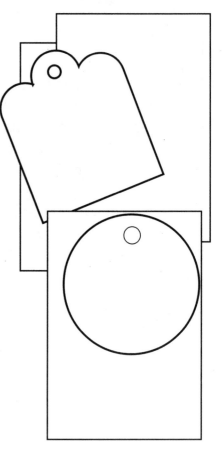

Construction:
1. Cut the pour spouts off of the milk cartons to make small boxes.
2. Cover the milk cartons with Contact paper.
3. Duplicate the price tags and cut out.
4. Attach one price tag to each small object and place each object in a milk carton.
5. Duplicate the money and coin patterns, laminate, and cut out.
6. Place the sets of play money in plastic bags for student use. Label one bag "buyer" and put in a ten dollar note. Label the other bag "seller" and put in a variety of change plus a five dollar note.

Directions for Two Students:
1. Decide which student will be the buyer and which the seller.
2. To buy the object in the first milk carton, the buyer gives the seller ten dollars.
3. The seller must figure out the correct change, count out the necessary money pieces, and put them in the milk carton with the item that was purchased.
4. The buyer should double-check his or her change.
5. Continue buying and selling the other items, switching roles as buyer and seller.

PRICE TAG PATTERNS

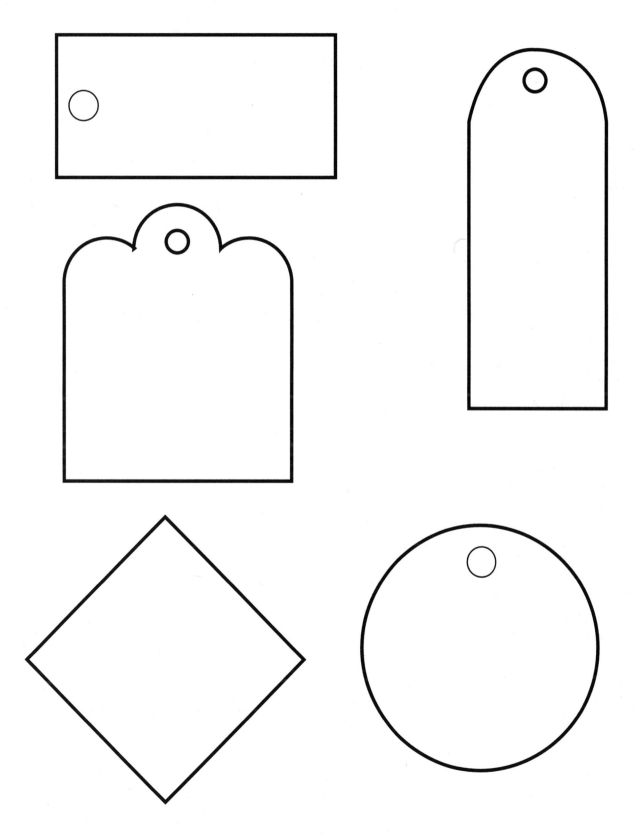

MONEY PATTERNS

CHAPTER FOUR

OLD MAGAZINES

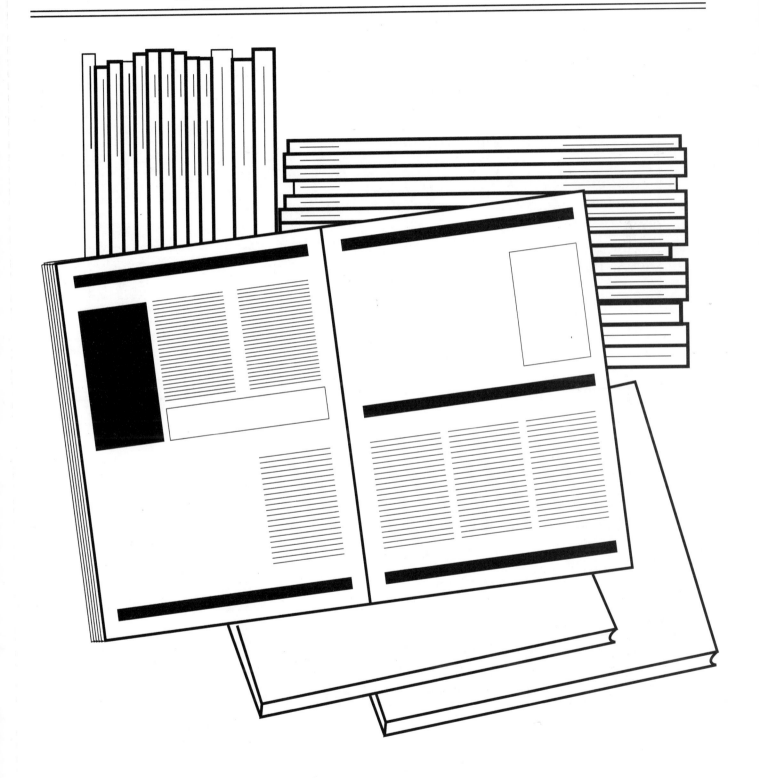

PICTURE HUNT

KINDERGARTEN

Objective: To practice size comparison skills.

Materials: magazines, large piece of <u>newsprint</u> or <u>butcher</u> paper, markers, glue, scissors, size patterns (p. 33)

Construction:
1. Fold the large paper in half, then unfold it and draw a line down the crease in the center.
2. Write the word "Smaller" on one side of the line and "Bigger" on the other.
3. On the center line—overlapping a bit onto both sides—glue one of the size patterns. Children will then cut pictures from the magazines and place them on one side of the size pattern (deciding whether the object pictured is bigger or smaller than the one in the center). You can extend this activity by changing the shape pattern.

Directions for Student:
1. Look through the magazines.
2. Cut out pictures of objects.
3. Glue pictures that are bigger than the picture in the middle of the page under the word "Bigger." Glue pictures that are smaller than the picture in the middle under the word "Smaller."

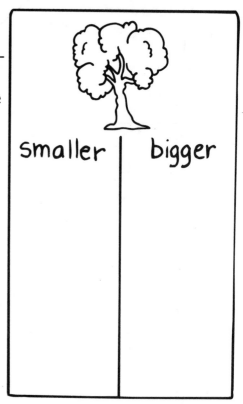

SIZE PATTERNS

SIZE PATTERNS

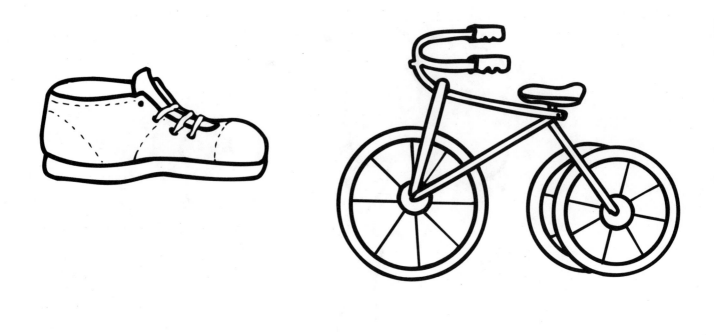

WHAT'S THE SALE PRICE?

FIRST GRADE

Objective: To reinforce subtraction skills.

Materials: magazines, scissors, 15 pieces of tagboard, glue, markers, plastic bag, paper, pencils

Construction:
1. Cut out 15 items from magazines. Choose items that children might be interested in buying.
2. Mount the pictures on the tagboard cards and number them on the back from 1 to 15.
3. Write a real price for each item on each tagboard card, for example, a pair of shoes might cost $33.
4. Write an amount saved during a sale beneath the real price, for example, a pair of shoes might be $12 off.
5. Laminate the cards and store in a plastic bag.
6. Make an answer key.

Directions for Student:
1. Number a piece of paper from 1 to 15.
2. Select a card and write a subtraction problem for it. For example, if the pair of shoes on the card regularly cost $33 and are $12 off, the problem would read: 33 - 12 =
3. Work out the problem for each card. Write the answer by the correct number on your paper.
4. Turn in your paper for correction.

1. $33 - 12 =$ ____
2. $44 - 10 =$ ____
3. ____
4. ____
5. ____
6. ____
7. ____
8. ____
9. ____
10. ____
11. ____
12. ____
13. ____
14. ____
15. ____

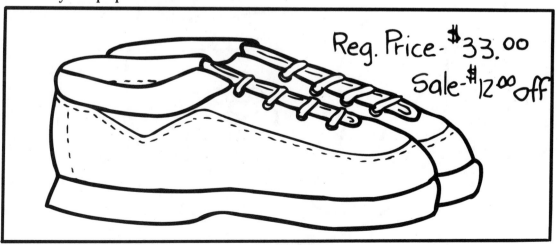

Reg. Price - $33.00
Sale - $12.00 off

35

CAFETERIA MEALS

SECOND GRADE

Objective: To reinforce calculator skills.

Materials: 10 pictures of plates of food cut from cooking magazines, 10 pieces of tagboard, markers, glue, empty box, calculators, paper, pencils

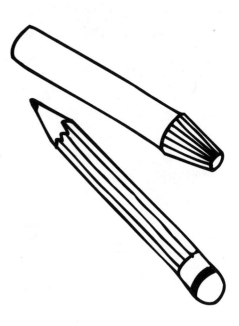

Construction:
1. Mount the pictures onto the tagboard cards.
2. Number the cards from 1 to 10.
3. Write the name of each item on the plate and a price for each item beside the picture. (See the illustration.)
4. Store the cards in an empty box.

Directions for Student:
1. Number a paper from 1 to 10.
2. Using a calculator, add up the cost of each meal. Then write the answer beside the correct number on your paper.
3. Turn in the paper for correction.

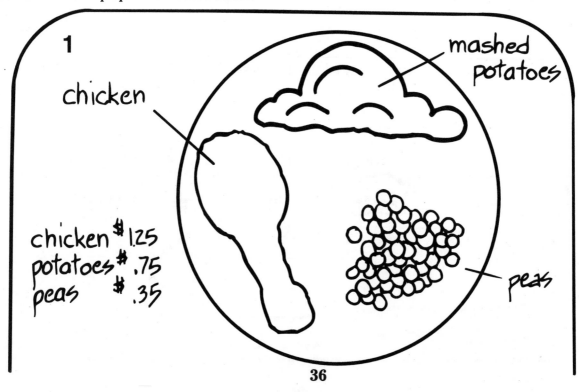

WHEN IS IT DONE?

THIRD GRADE

Objective: To reinforce time-telling skills.

Materials: 10 recipes cut from cooking magazines (the recipes must have an exact cooking or baking time), 10 pieces of tagboard, marker, clock patterns (p. 38), wax crayons, glue, scissors, plastic bag

Construction:
1. Mount each recipe on a piece of tagboard.
2. Duplicate the clock patterns and cut them out.
3. Glue two clocks under each recipe.
4. Draw the hands on the first clock to show when the item being cooked or baked was put in the oven.
5. Laminate the activity cards and store in a plastic bag with a wax crayon.

Directions for Student:
1. Read the recipe to find out how long the item is to be cooked or baked.
2. Look at the hands on the first clock to find the time that the item was placed in the oven.
3. Figure out what time the item will be done cooking.
4. Use a wax crayon to draw the ending time on the second clock.
5. Turn your work in to the teacher for correction.

Chocolate Chip Cookies

1 cup flour
1 cup sugar
1 egg
$1/2$ cup butter
$1/2$ cup chocolate chips
$1/2$ teaspoon salt
$1/2$ teaspoon vanilla
$1/2$ teaspoon baking soda

Mix all ingredients together and drop by rounded teaspoonfuls onto greased cookie sheet. Bake 12 minutes at 350° or until done.

CLOCK PATTERNS

CHAPTER FIVE

CONSTRUCTION PAPER SCRAPS

MARVELOUS MOBILES

KINDERGARTEN

Objective: To reinforce shape concepts.

Materials: construction paper scraps, large shape patterns (p. 41), small shape patterns (p. 42), paper plates (1 per child), tagboard, scissors, yarn, glue, hole punch, tape or stapler

Construction:
1. Punch four holes at equal intervals around each paper plate.
2. Tie a 6" length of yarn from each hole.
3. Tape or staple a 12" piece of yarn from the middle of each plate for hanging purposes.
4. Duplicate the shape patterns onto tagboard and cut out to make stencils. Make enough for student use.
5. After the children make their shapes, help them tie them to the pieces of yarn.
6. Hang the completed mobiles from the ceiling or from a clothesline strung across the classroom.

Directions for Student:
1. Use the stencils to copy the shape patterns onto construction paper scraps. Cut each out.
2. Glue each small shape onto the center of the matching large shape.
3. Punch a hole in the top of each large shape.
4. Tie one shape to the end of each of the four short pieces of yarn on the paper plate.

LARGE SHAPE PATTERNS

SMALL SHAPE PATTERNS

SHE'LL BE WEARING RED PAJAMAS!

FIRST GRADE

Objective: To practice graphing skills.

Materials: construction paper scraps, pen, game markers (p. 44), clothing patterns (p. 45), scissors, large sheet of butcher paper

Construction:
1. Duplicate the clothing patterns onto various colors of construction paper scraps, and cut them out.
2. Make a six-column chart on the piece of butcher paper. Glue one of the clothing patterns to the top of each column, for example, blue jeans in the first column, a white T-shirt in the second, a pink dress in the third, a green coat in the fourth, white sneakers in the fifth, and red pj's in the sixth.
3. Place the chart on a table.
4. Have the students take turns placing a marker in the columns that fit what they are wearing. (The children will definitely laugh at the thought of someone wearing pajamas to school!)
5. You can make different charts by copying the patterns onto different colors of construction paper. Then have children compare the results from one day to another.

Directions for Student:
1. Take a turn placing a marker in each column that fits what you are wearing.
2. Count up how many markers are in each column.

43

MARKERS

CLOTHING PATTERNS

PERFECT PIZZAS

SECOND GRADE

Objective: To master working with fractions.

Materials: construction paper scraps, tagboard, manila drawing paper, pizza topping patterns (p. 47), large circle pattern (p. 48), scissors, pencils, markers, glue, scissors, paper scraps

Construction:
1. Duplicate the topping patterns and circle pattern onto tagboard and cut out to make stencils.
2. Show the children how to divide their papers into four sections by folding the paper horizontally and then vertically.
3. Discuss with your students the different ways to divide a pizza, and explain that when they take one slice they are taking a fraction of the pizza.

Directions for Student:
1. Trace around one of the circular stencils to make a pizza shape in each section of your paper (four total).
2. Write "1/2" in the first section, "2/3" in the second section, "3/4" in the third section, and "5/6" in the fourth section.
3. Copy the pizza topping stencils onto construction paper scraps and cut out. You can make shredded cheese by cutting up construction paper.
4. Cover the fractional amount of the pizza that is listed in each section with paper scrap toppings.

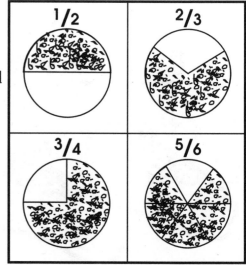

PIZZA TOPPINGS

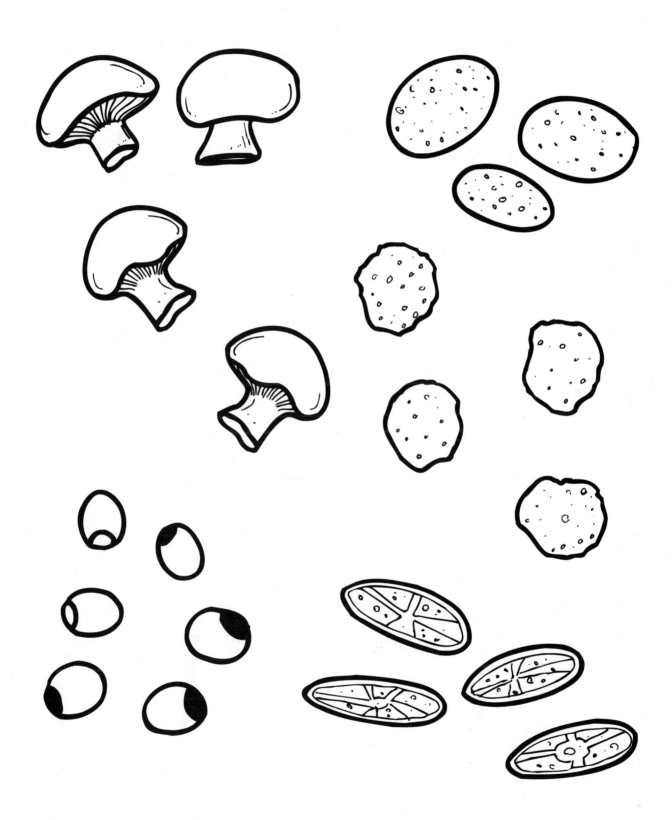

CIRCLE PATTERNS

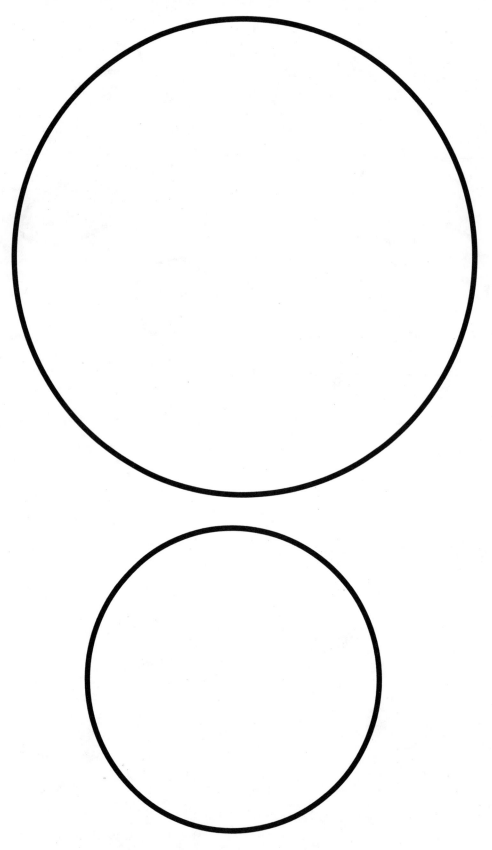

THE PERIMETER, PLEASE!

THIRD GRADE

Objective: To reinforce ruler and addition skills in inches.

Materials: construction paper, shape patterns (p. 50), scissors, plastic bag, marker, pencils, paper, rulers

Construction:
1. Copy the shape patterns onto construction paper and laminate.
2. Number the back of each shape from 1 to 8.
3. Cut out the shapes and store in a plastic bag.

Directions for Student:
1. Number your paper from 1 to 8.
2. Measure each side of shape #1 and add the numbers together to get the perimeter. Write down the number on your paper.
3. Continue measuring each shape, writing down the perimeter on your paper.
4. Turn in your paper for the teacher to check.

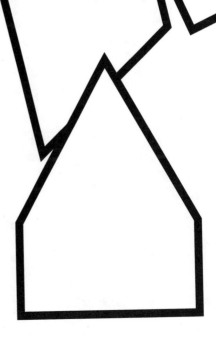

SHAPE PATTERNS

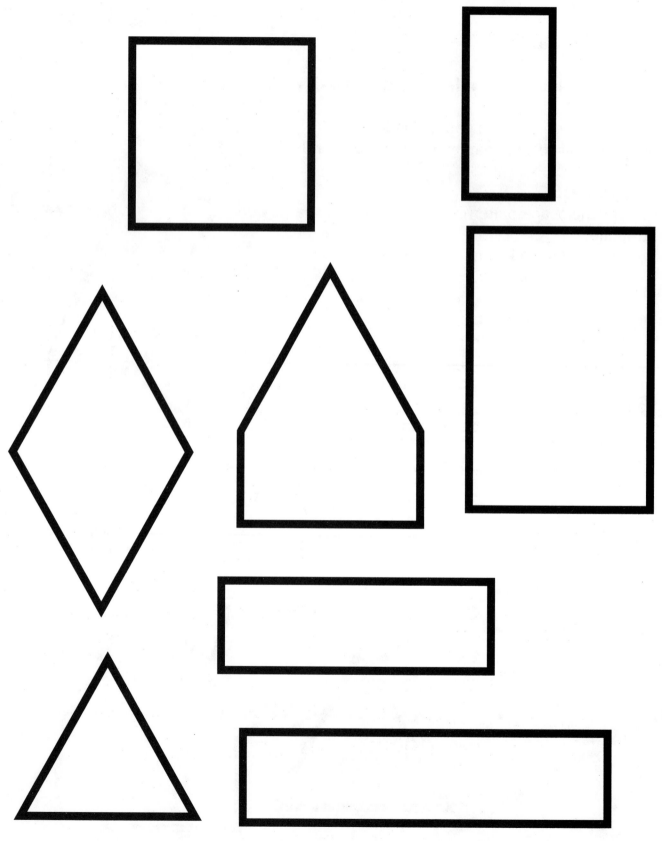

CHAPTER SIX

PAPER BAGS

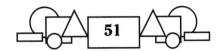

CATERPILLAR MATH

KINDERGARTEN

Objective: To reinforce number sequencing.

Materials: paper bags cut open so they lay flat (1 per child), circle patterns (p. 48), tagboard, markers, sets of small items (popcorn, buttons, beans, pebbles, shells), glue, scissors

Construction:
1. Make circle stencils by copying both circle patterns onto tagboard and cutting out.
2. Trace a large caterpillar head in the lower left-hand corner of each flat paper sack (see the illustration).
3. Make a model project (see directions for the student) for children to follow.
4. Help the children number their caterpillar body segments from 1 to 10.

Directions for Student:
1. Trace the smaller circle stencil to make 10 caterpillar body parts. Attach in a row to the caterpillar head.
2. With the teacher's help, number the body parts from 1 to 10.
3. Glue the right number of objects onto each part to match its number. For example, glue one pebble in the first part, two buttons in the second, and so on.

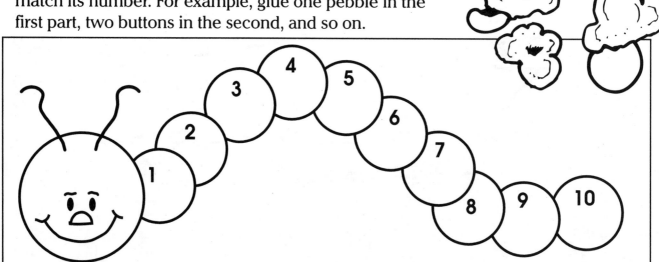

52

CLIMBING A MATH LADDER

FIRST GRADE

Objective: To reinforce ordinal word recognition.

Materials: paper bags cut open so they lay flat (1 per child), short strips of construction paper (10 per child), markers, glue, crayons, rulers, completed pattern (p. 54)

Construction:
1. Write the ordinal words on the board for children to copy or use to check their work.
2. Enlarge the pattern to put up as a model for the children to follow.

Directions for Student:
1. Copy the ordinal words onto the construction paper strips.
2. Use a marker and a ruler to make two long vertical marks a few inches apart on your paper. (These will serve as the sides of your ladder.)
3. Glue the construction paper strips in the correct order up the ladder (for rungs), beginning with "first" and ending with "tenth."
4. If you'd like, draw a picture around the ladder, for example, a tree house, an apple tree, or a fire truck.

tenth

ninth

eighth

seventh

sixth

fifth

fourth

third

second

first

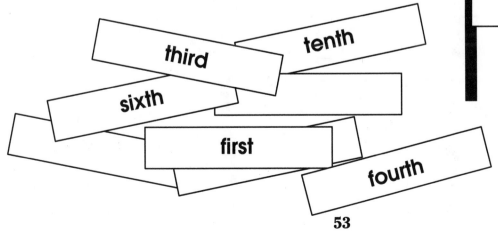

third

tenth

sixth

first

fourth

LADDER PATTERN

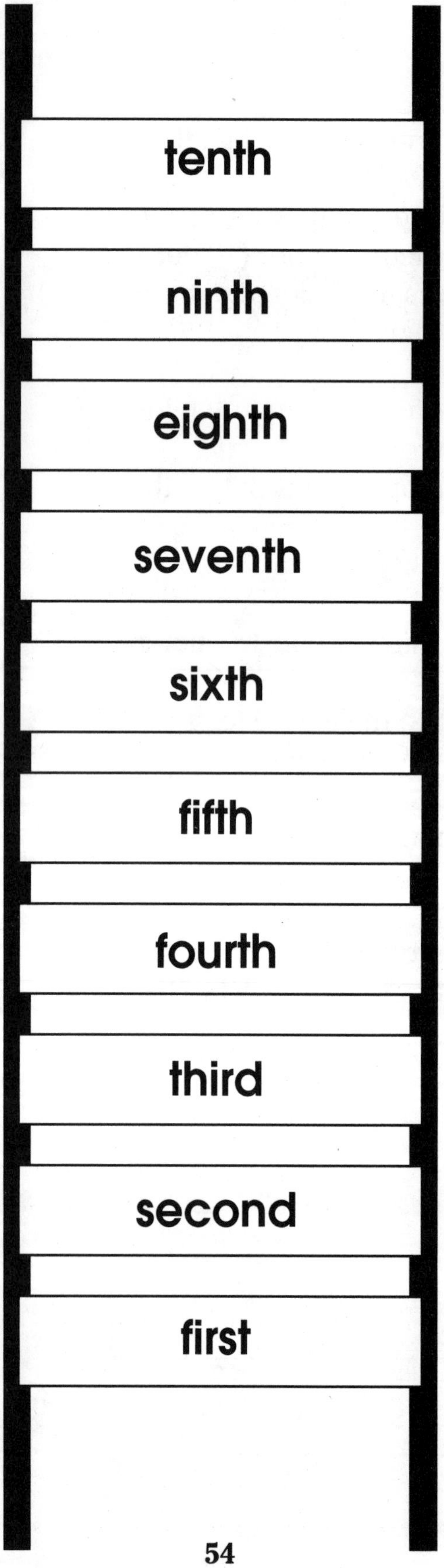

tenth

ninth

eighth

seventh

sixth

fifth

fourth

third

second

first

PIGS IN A PEN

SECOND GRADE

Objective: To reinforce addition skills using all types of coins.

Materials: paper bag, scissors, markers, coin patterns (pp. 26 and 56), pig patterns (p. 57), glue, empty box

Construction:
1. Cut the paper bag along the seams so it lays flat.
2. Make a 15-section grid ("pigpen") on the paper sack.
3. Write a random money amount in each section.
4. Duplicate 15 pigs and cut them out.
5. Glue sets of coins onto each pig so that the coins equal the money amounts in the pigpen squares.
6. Laminate the pigs and cut them out.
7. Store the pieces in the box.

Directions for Student:
1. Spread the paper bag so that it lays flat on a table or desk.
2. Figure out the amount of money on each pig and place that pig in the correct section of the pigpen.
3. Show your teacher the pigpen for correction when you've finished.

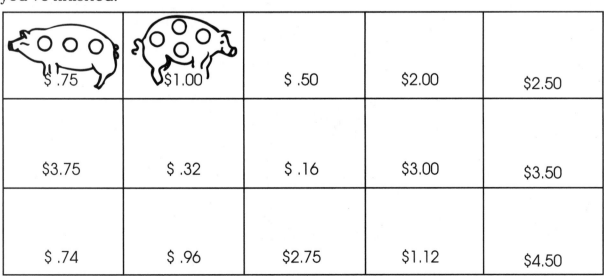

$ .75	$1.00	$ .50	$2.00	$2.50
$3.75	$ .32	$ .16	$3.00	$3.50
$ .74	$ .96	$2.75	$1.12	$4.50

COIN PATTERNS

PIG PATTERNS

FROM MY HOUSE TO SCHOOL

THIRD GRADE

Objective: To reinforce mapping skills.

Materials: paper bags (1 per child), crayons or markers, scissors

Construction:
1. Show the students how to cut the paper bag down the seams to make a flat working surface.
2. Make a model map for the students showing how you get home from school.

Directions for Student:
1. Draw a map that shows how you get home from school.
2. Be sure to add as much detail as possible, including local landmarks, parks, and important buildings.

CHAPTER SEVEN

EMPTY BOXES

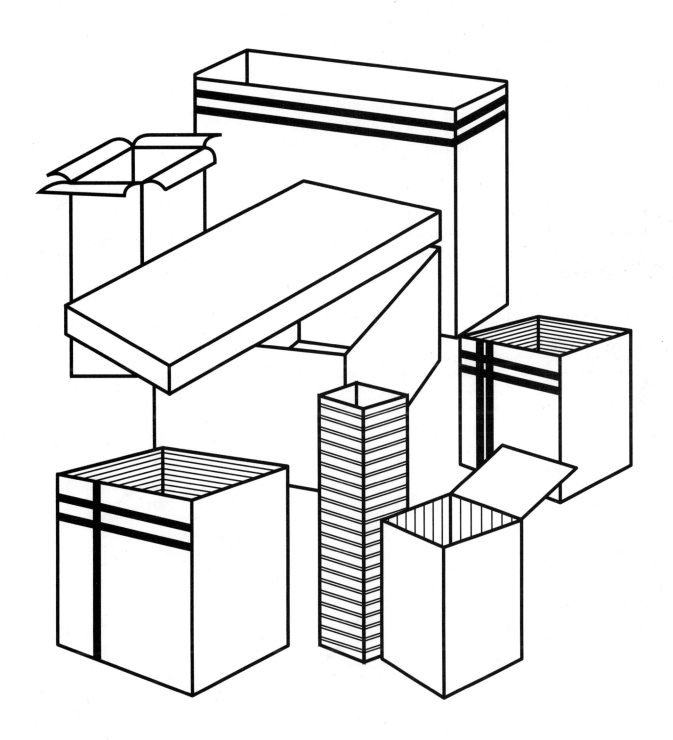

BOX PAIRS

KINDERGARTEN

Objective: To reinforce matching skills.

Materials: empty box, assorted pairs of objects (buttons, paper clips, barrettes, rocks, feathers, acorns, pine cones, wooden beads, plastic spoons, blocks, erasers, etc.), construction paper, colored Contact paper

Construction:
1. Cover the box with Contact paper to make it durable.
2. Place 10-12 pairs of items in the box.

Directions for Student:
1. Put a piece of construction paper on your desk.
2. Empty the box onto the paper.
3. Match the items into pairs.

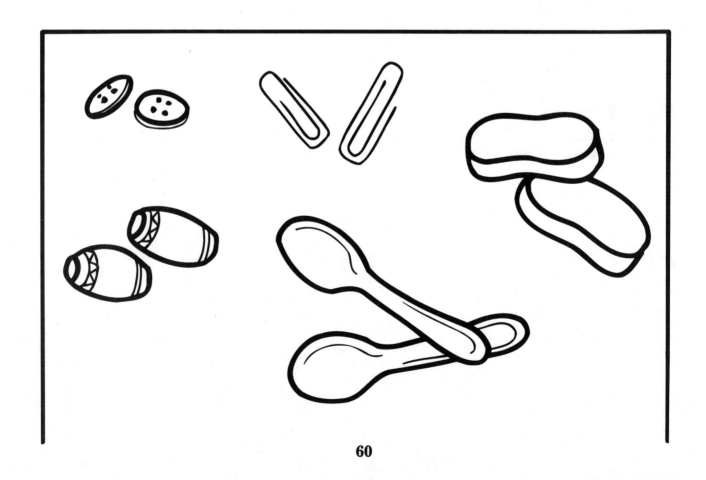

60

HOW MANY PENNIES?

FIRST GRADE

Objective: To reinforce estimation skills.

Materials: empty box with lid, jar of jelly beans or pennies, rubber bands or tape, score sheets (p. 62; 1 per student), paper, pencils

Construction:
1. Place 10 to 20 jelly beans or pennies in the box. (After each round of guessing, place more objects into the box.)
2. Close the box and secure it with rubber bands or tape so that it won't open when shaken.
3. Pass the box around the class, asking each child to write down a guess as to how many items are in the box. Have the students keep their guesses a secret.
4. When all of the students have guessed, make a graph of the children's estimates, listing how many students guessed each number.
5. Open the box and count the items to see which students were closest. Let the children keep track of their guesses on their score sheet.

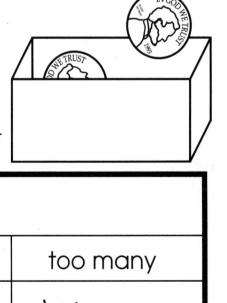

Directions for Student:
1. Shake the box and write down your estimate of how many items are inside.
2. When the items have been counted, mark on your paper whether you guessed not enough, exactly right, or too many.

I guessed		
not enough	exactly right	too many
||	|||	＋＋＋

SCORE SHEET

I guessed		
not enough	exactly right	too many

HOW FAR DID THE BOX TRAVEL?

SECOND GRADE

Objective: To practice map skills and measuring skills.

Materials: 15 or more food containers or household products, large map of the United States mounted on the bulletin board, lengths of yarn, yardsticks and rulers, calculator, scissors

Construction:
No preparation is needed for this activity.

Directions for Student:
1. Select a box and read the print to find out where the item was manufactured. Find that state on the map.
2. Place one end of a length of yarn in your own state and the other in the state where the product was manufactured.
3. Cut the yarn to match the distance and then measure it.
4. Use the map key to find out how many inches equal a mile, and then figure out how many miles the length of yarn is equal to.
5. Continue finding the distance for each box.

Option: This activity can also be done by pairs of students.

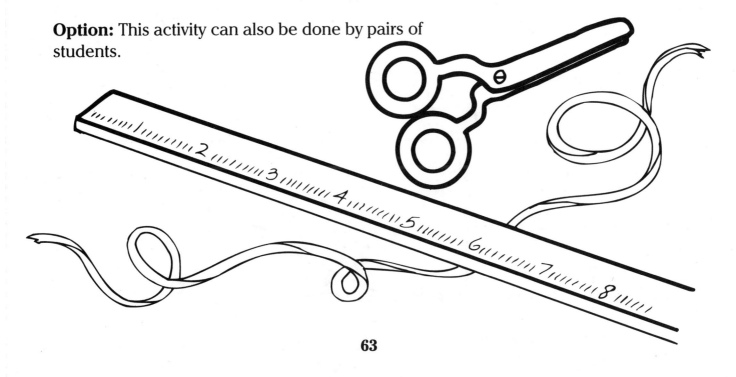

MATH RACES

THIRD GRADE

Objective: To practice addition and subtraction.

Materials: small boxes (1 box for every 2 children), number strips (p. 65; 1 set for each box), tagboard, calculators, paper, pencils, scissors

Construction:
1. Duplicate the number strips pattern page onto tagboard, making one set for each box.
2. Laminate the numbers and cut them out.
3. Store a set of the number strips in each box.

Directions for Two Students:
1. Select a box of number cards.
2. Take turns pulling out two number cards at a time.
3. Write an addition and a subtraction problem for the two numbers.
4. Try to solve both problems. The student who completes both problems correctly first wins a point. (Check answers with a calculator.)
5. Continue playing until all cards are used. The student with the most points wins.

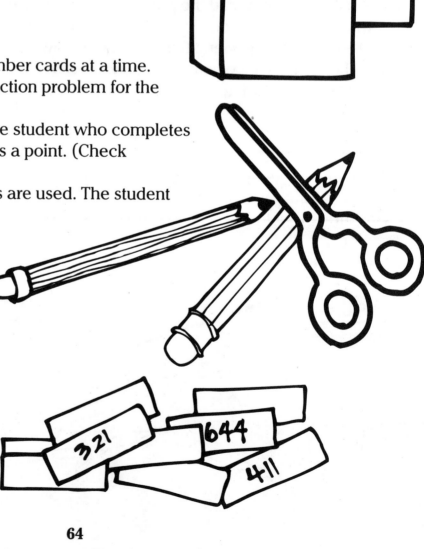

NUMBER STRIPS

123

321

742

930

217

411

NUMBER STRIPS

329

670

542

888

721

644

OUTDATED CALENDARS

CALENDAR MATH

KINDERGARTEN

Objective: To reinforce number sequencing.

Materials: 1 page of calendar numbers for each child, construction paper, scissors, glue

Construction:
1. Depending on your students' abilities, cut the calendar pages to include numbers from 1 to 10 or 1 to 25.
2. Cut the construction paper into long strips.

Directions for Student:
1. Cut apart the page of numbers.
2. Glue each number onto the construction paper strip in the correct sequence.

21

13

11

31

12

6

4

19

23

8

30

18

THURSDAY + SATURDAY = ?

FIRST GRADE

Objective: To reinforce basic math facts.

Materials: 1 page of calendar numbers for each child, manila drawing paper, scissors, glue

Construction:
No preparation is required for this activity.

Directions for Student:
1. Cut apart your page of numbers.
2. See how many math facts you can make out of the numbers, for example,   2 + 4 = 6, 24 - 11 = 13, 5 + 9 = 14
3. Glue the problems onto the paper as you make them.
4. Give any extra numbers to the teacher to use in another activity.

$$12 + 4 = 16$$

$$24 - 11 = 13$$

STORY PROBLEMS

SECOND GRADE

Objective: To practice writing and working story problems.

Materials: old calendar with large numbers, lunch sack, 3" x 5" index cards, scissors, pencils, paper

Construction:
1. Cut out all of the numbers on several calendar pages (save the pictures for use in another activity).
2. Place the numbers in the lunch sack for student use.
3. Give each student a 3" x 5" index card.
4. When all of the story problems are complete, have each child read his or her problem out loud to the class while the other students try to solve it.

Directions for Student:
1. Pick two numbers from the lunch sack.
2. Write a story problem using the two chosen numbers.

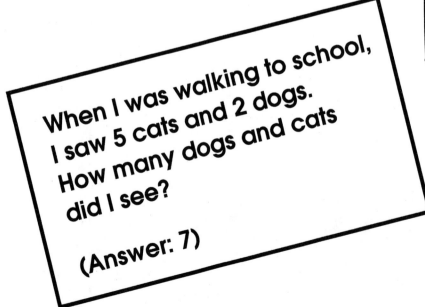

When I was walking to school, I saw 5 cats and 2 dogs. How many dogs and cats did I see?

(Answer: 7)

CALCULATOR ADDITION

THIRD GRADE

Objective: To practice adding with a calculator.

Materials: old calendar pages (1 per child), number page (p. 72), manila paper, scissors, glue, pencils, calculators

Construction:
1. Duplicate the number page.
2. If children run out of numbers to use from their calendar pages, they may use the extra numbers on the number page.

Directions for Student:
1. Cut apart the date squares on the calendar page.
2. Make addition problems, using four of the date squares for each problem.
3. Glue the numbers in a row on a piece of manila paper, making sure to place the addition sign in the correct spot.
4. Add each set of numbers in your head.
5. Use the calculator to check your answers.
6. Trade papers with another student for correction.

$$
\begin{array}{r}
4 \\
5 \\
16 \\
+\ 20 \\
\hline
45
\end{array}
$$

1 13 17 2

NUMBERS

1	2	3	4	5
6	7	8	9	10
11	12	13	14	15
16	17	18	19	20
21	22	23	24	25
26	27	28	29	30
31				

CHAPTER NINE

OLD NEWSPAPERS

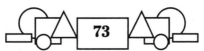

WHICH COSTS MORE?

KINDERGARTEN

Objective: To reinforce comparison skills.

Materials: newspaper advertisements, 10 tagboard cards (3" x 5"), scissors, glue, marker, paper, pencils, plastic bag

Construction:
1. Number the tagboard cards from 1 to 10.
2. Find items in the newspaper advertisements with large, readable prices, and glue two on each of the cards.
3. Label the first item on each card "A" and the second item "B."

Directions for Student:
1. Number your paper from 1 to 10.
2. Look at each card and decide which item is more expensive.
3. Write down the correct letter (either "A" or "B") next to the matching number on your paper.
4. Continue working until all 10 cards are completed.

THE PRICE IS RIGHT!

FIRST GRADE

Objective: To reinforce number sequencing.

Materials: newspaper advertisements, 5 different colors of construction paper (cut each into five 6" x 6" squares), glue, plastic bag, scissors

Construction:
1. Cut out 25 items and their prices from old newspapers.
2. Arrange the items and prices into sets of five, making sure that all items in any given set have different prices.
3. Glue the sets onto the different-colored squares of construction paper—you will have five items glued on five pieces of red construction paper, five items glued on five pieces of green construction paper, and so on.
4. Laminate all of the cards and cut them out.
5. Store all pieces in a plastic bag.

Directions for Student:
1. Sort the activity pieces into sets by color.
2. Arrange the pieces in each set from the least to the most expensive.
3. Show your work to the teacher before putting the squares back in the plastic bag.

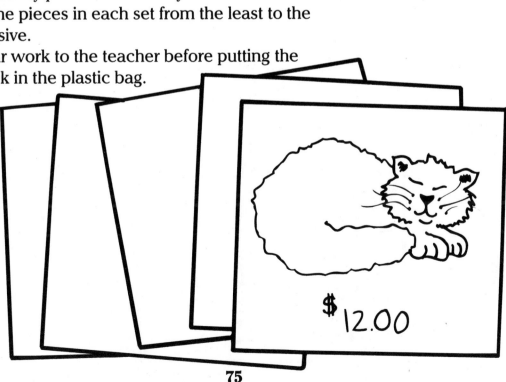

GROCERY MONEY

SECOND GRADE

Objective: To reinforce money concepts.

Materials: grocery advertisements, tagboard, scissors, dollar patterns (pp. 30 and 77), coin patterns (pp. 26 and 56), plastic bag, glue

Construction:
1. Cut out 15 to 20 pictures of food items and their prices. (The prices should cover a wide range of amounts.)
2. Glue the pictures onto individual pieces of tagboard, laminate, and cut out.
3. Duplicate the dollar and coin patterns, laminate, and cut out.
4. Store all pieces in a plastic bag.

Directions for Student:
1. Take out each picture and look at the price.
2. Lay out the correct amount of money that you would need to buy the item.
3. Continue with each card until you have used all of your grocery money or gone through all of the cards.
4. Choose which items you really like if you do not have enough money to buy everything.

Cookies $1.35

DOLLAR PATTERNS

PHONE NUMBER MATH

THIRD GRADE

Objective: To practice number sequencing in the thousands.

Materials: newspaper classified sections, scissors, glue, paper

Construction:
No preparation is required for this activity.

Directions for Student:
1. Look through the newspaper page and cut out 10 phone numbers.
2. Cut off and keep the last four digits in each number, and throw the rest away.
3. Paste the 10 four-digit numbers on your paper in order from smallest to largest number.

1099
1155
1250
1255
3244
3373
4488
6875
8875
9910

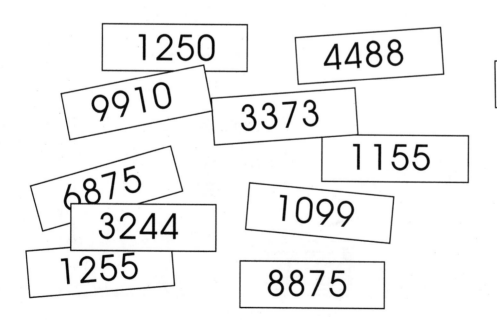

CHAPTER TEN

WRAPPING PAPER

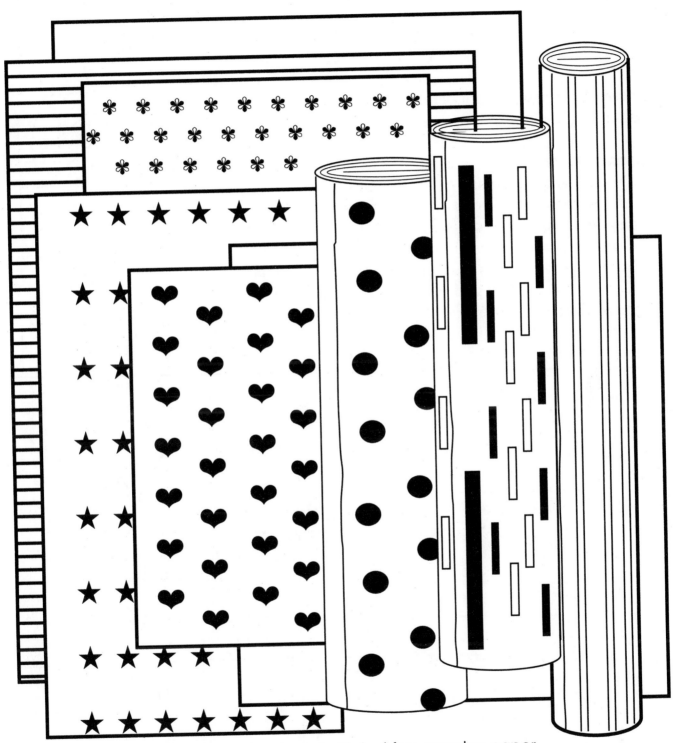

Wallpaper may be substituted for wrapping paper.

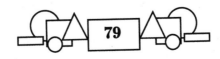

SORTING BY SHAPE

KINDERGARTEN

Objective: To reinforce size and shape sorting.

Materials: an assortment of decorated wrapping paper, plastic bag, tagboard, shape grid (p. 81), scissors, glue

Construction:
1. Cut the wrapping paper to match the shapes on the chart below; make several of each size shape. Cut all of the same shapes from the same patterned paper.
2. Glue the shapes onto tagboard, laminate, and cut out.
3. Scramble the shapes and store them in a plastic bag.
4. Duplicate the shape grid onto tagboard and laminate for student use.

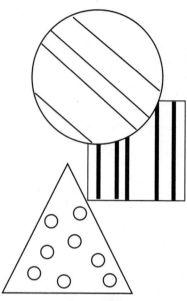

Directions for Student:
1. Sort the wrapping paper by shapes.
2. Place all of the big circles on the big circle area of the shape grid, all of the medium-sized circles on the medium-sized circle area of the grid, and so on. (See the illustration.)
3. Show the finished grid to the teacher before putting the shapes away.

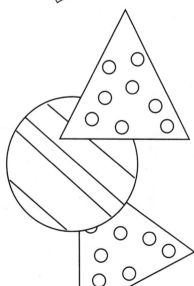

Shape Grid

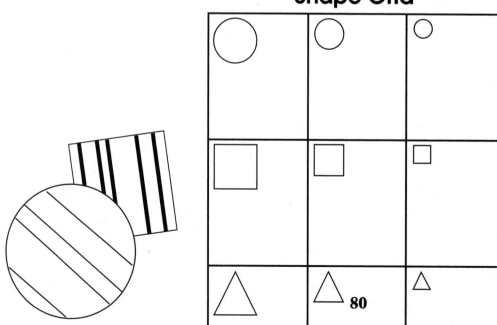

80

SHAPE GRID

PRESENTING . . . MATH!

FIRST GRADE

Objective: To reinforce ones and tens places.

Materials: wrapping paper, 20 small boxes or milk cartons, number cards (p. 83), tape, scissors, glue

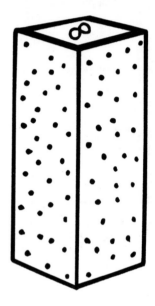

Construction:
1. Wrap the small boxes or milk cartons to look like presents.
2. Make two copies of the number cards and glue one to the top of each present.
3. Say a two-digit number and then call on a student to pick out two cartons to represent the tens and the ones places of the number called.

Directions for Student:
1. When the teacher calls a number, think about which two boxes would make up that number.
2. If the teacher calls on you, pick up the two boxes that represent the tens and ones places of the number.

NUMBER CARDS

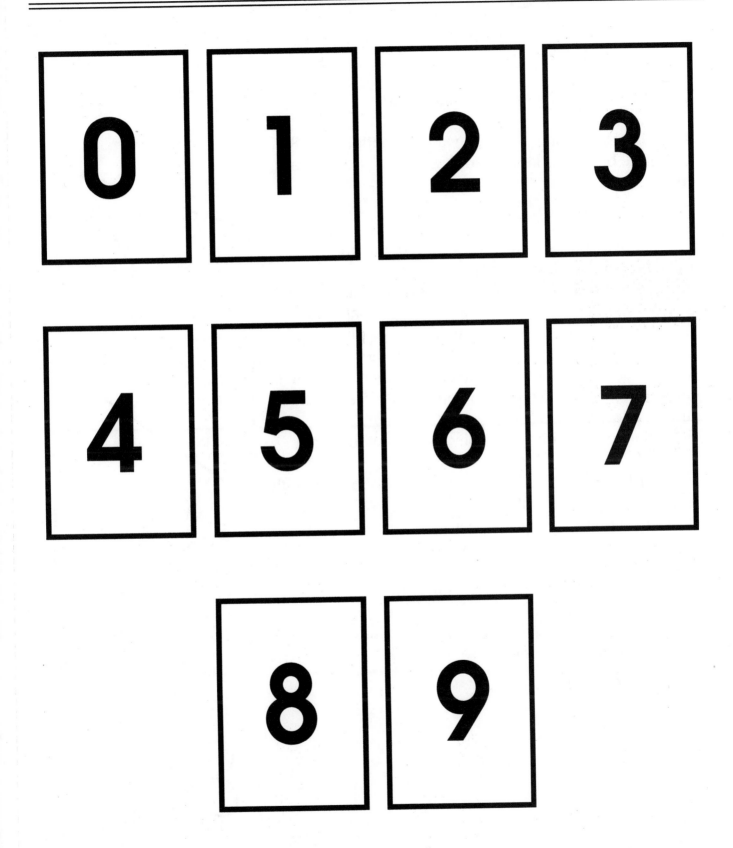

WHAT'S THE MISSING SIGN?

SECOND GRADE

Objective: To reinforce basic math facts.

Materials: wrapping paper, kite and tail patterns (p. 85), tagboard, black marker, scissors, glue, hole punch, yarn, markers, plastic bag

Construction:
1. Make two copies of the kite pattern on tagboard, cut out, and cover with wrapping paper.
2. Make 30 copies of the tail pattern on tagboard, cut out, and cover with a variety of wrapping paper patterns.
3. Use a thick, black marker to write a large plus (+) sign on one kite and a large minus (-) sign on the other kite.
4. Write one math fact on each kite tail piece, omitting the plus or minus sign in each problem. Make 15 subtraction and 15 addition facts.
5. Laminate all of the activity pieces and cut them out.
6. Punch two holes in each kite tail piece.
7. Attach to each kite a length of yarn that is long enough to lace 15 tail pieces onto it.
8. Store all pieces in a plastic bag.

Directions for Student:
1. Take out the two kites and the tail pieces.
2. Look at the math fact on each tail piece and decide whether there is a plus or minus sign missing.
3. Lace the tail sections onto the correct kite.

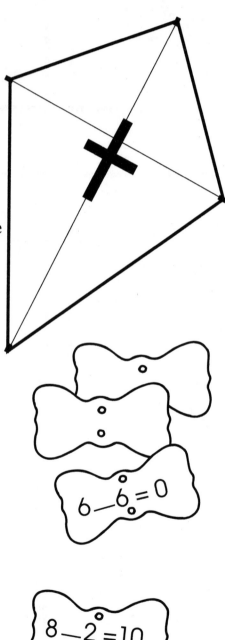

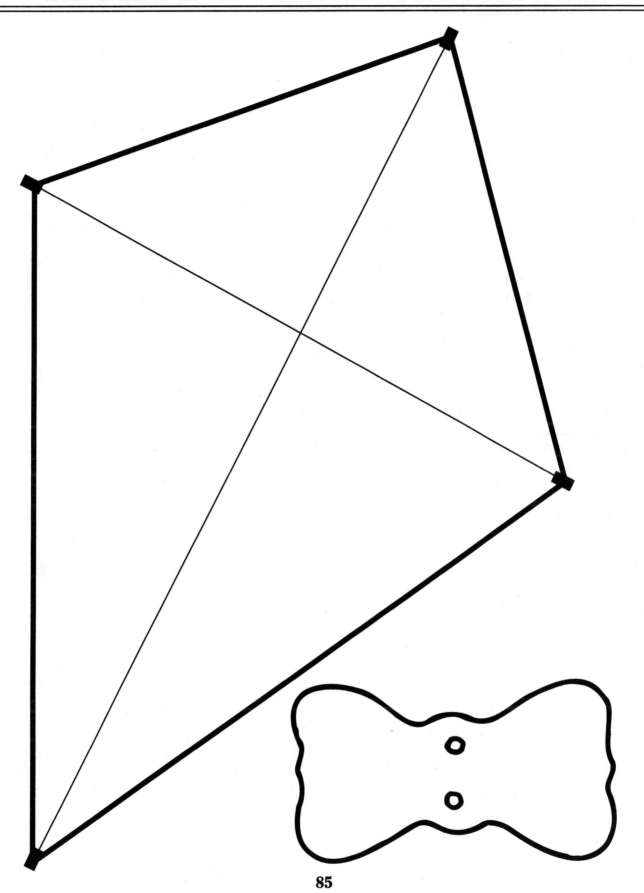

FRACTION PUZZLES

THIRD GRADE

Objective: To reinforce knowledge of fractions.

Materials: wrapping paper, shape patterns (pp. 87-88), tagboard, scissors, marker, wrapping paper, scissors, glue, plastic bag, paper, pencils

Construction:
1. Copy the 10 shapes onto tagboard and number the back of each from 1 to 10.
2. Cover a fraction of each shape with wrapping paper cut to fit the chosen area.
3. Laminate the shapes and store them in a plastic bag.

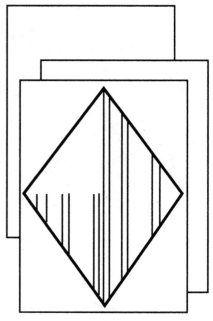

Directions for Student:
1. Number your paper from 1 to 10.
2. Empty the shapes onto a flat surface.
3. Look at the part of each shape that is covered with wrapping paper.
4. Write the correct fraction beside the matching number on your answer sheet.

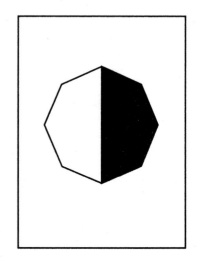

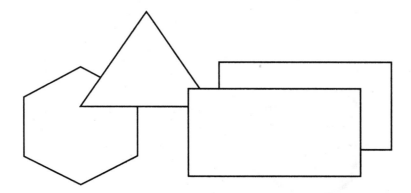

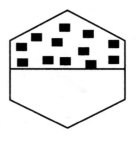

SHAPE PATTERNS

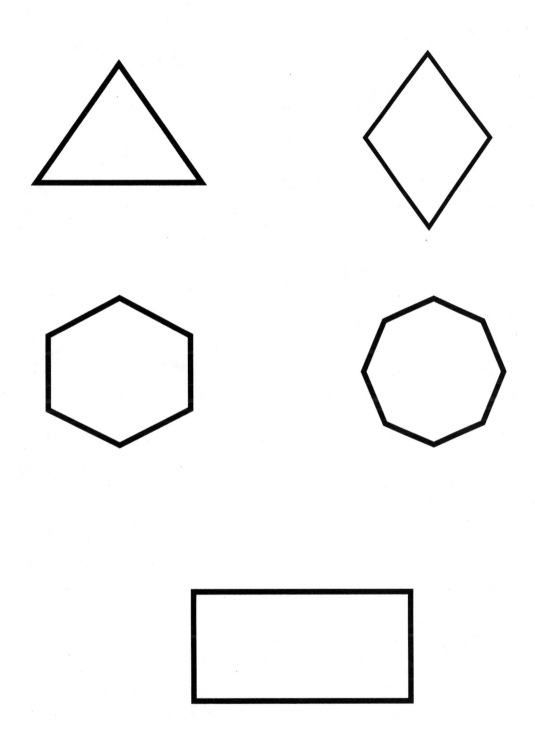

SHAPE PATTERNS

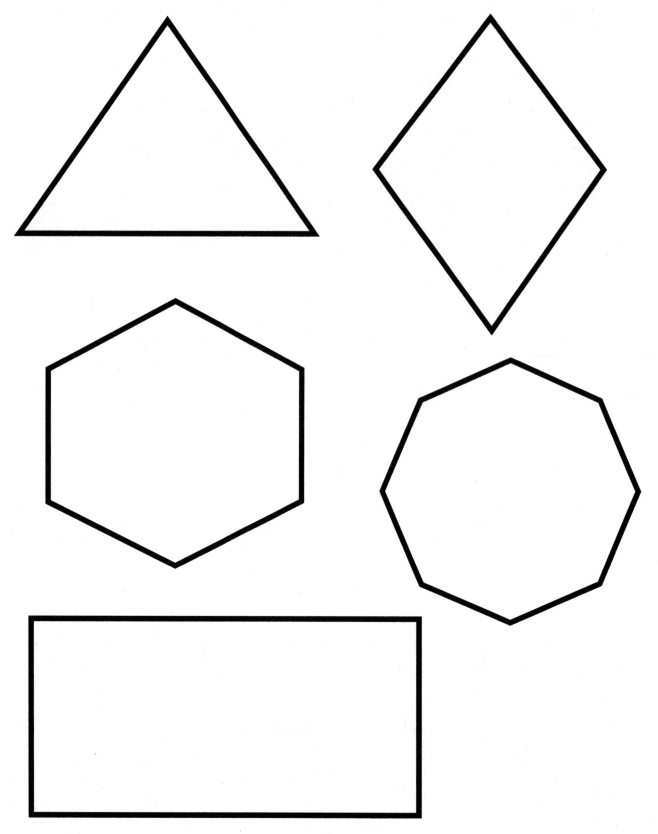

CHAPTER ELEVEN

POPSICLE STICKS

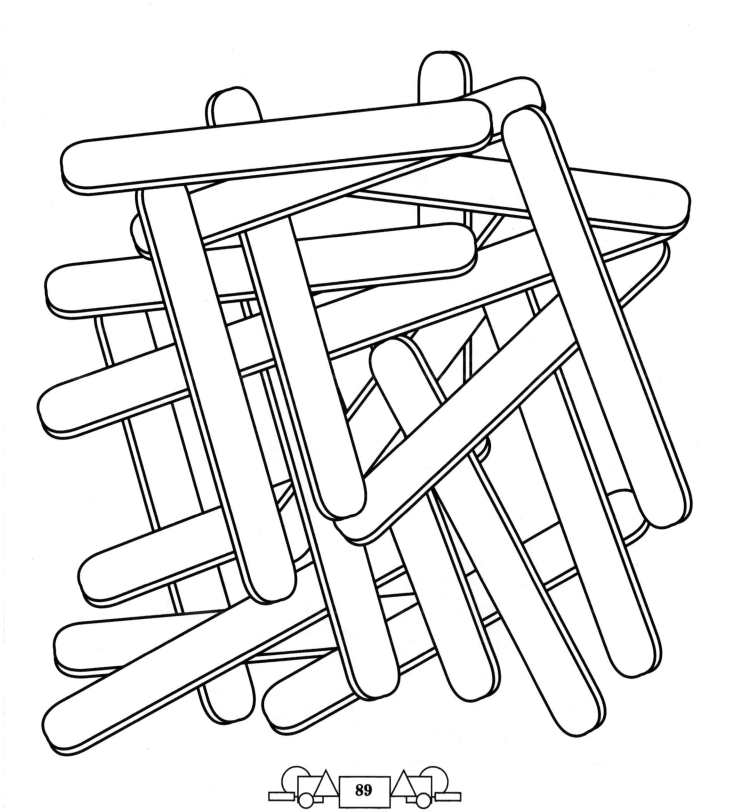

I AM SO BIG!

KINDERGARTEN

Objective: To reinforce measuring concepts.

Materials: Popsicle sticks, marker, oak tag, pens

Construction:
1. Draw a graph on a large sheet of oak tag similar to the one shown below. (Note: Most kindergarten students will range from 9 to 15 Popsicle sticks tall, so you'll need about that many squares per column on your chart.)
2. Have the students get into pairs to do this activity.

Directions for Two Students:
1. Take turns doing this activity. Lie down on the floor while your partner measures you by placing Popsicle sticks next to you end to end to represent your height. Then measure your partner in the same way.
2. Go to the chart and mark the correct number of spaces to represent how many "sticks" tall you both are.

Option: Post the completed chart on a bulletin board for children to admire and observe. Do this activity again later in the year and compare the differences.

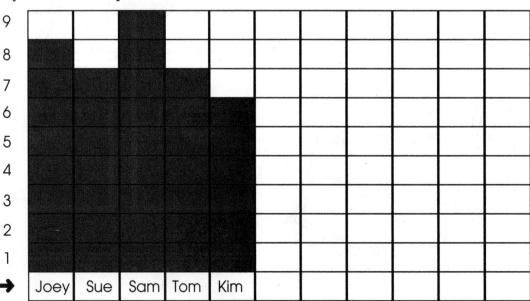

ADDITION OR SUBTRACTION?

FIRST GRADE

Objective: To reinforce mental math skills.

Materials: 3 Popsicle sticks for each child, glue gun and glue sticks, empty box

Construction:
1. Glue two sticks together to make a plus sign—make one for every child. The remainder of the sticks will serve as minus signs.
2. Give each child a plus and a minus sign.
3. Call out math facts with the plus or minus omitted.
4. Store the sticks in a box when not in use.

Directions for Student:
1. Listen as the teacher calls out math facts.
2. Hold up either the plus or minus sign to show which sign was left out of each math fact.

POPSICLE MEASURING

SECOND GRADE

Objective: To reinforce estimation skills.

Materials: 100+ Popsicle sticks, storage container, worksheet (p. 93; 1 per child)

Construction:
No preparation is required for this activity.

Directions for Student:
1. Estimate how many Popsicle sticks it will take to measure the length, width, or height of the items listed on the worksheet.
2. Write your estimate in the correct column.
3. Measure using Popsicle sticks and record how many sticks were actually used.
4. Compare your estimates with the actual measurements.

Name: _____		
Name of item to be measured	Estimate	Actual
Chalkboard - length		
Teacher's desk - height from floor		
Reading table - width		
Bookcase - height		
Sink - length		
Counter - length		

WORKSHEET

Name: _____

Name of item to be measured	Estimate	Actual
Chalkboard - length		
Teacher's desk - height from floor		
Reading table - width		
Bookcase - height		
Sink - length		
Counter - length		

POPSICLE PERIMETERS

THIRD GRADE

Objective: To reinforce measuring skills.

Materials: Popsicle sticks (6 per child), newsprint, rulers, markers or crayons

Construction:
No preparation is required for this activity.

Directions for Student:
1. Use four Popsicle sticks to make a shape on a piece of newsprint.
2. Draw around the Popsicle sticks and measure the outline of the shape to find the perimeter.
3. Arrange the four sticks to make a different shape.
4. Draw around the shape and measure its perimeter. (It should be the same length as the first shape.)
5. Follow the same directions using five Popsicle sticks to make two shapes and then six Popsicle sticks to make two shapes.

Note: No matter what shape students create, the perimeters will be the same if they use the same number of sticks!

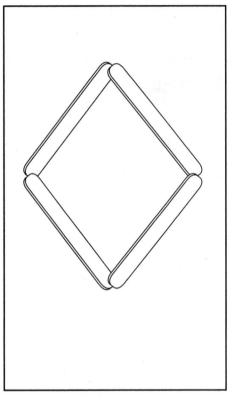

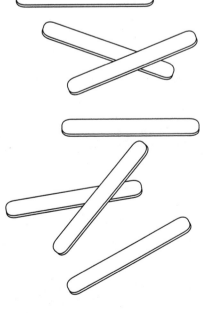

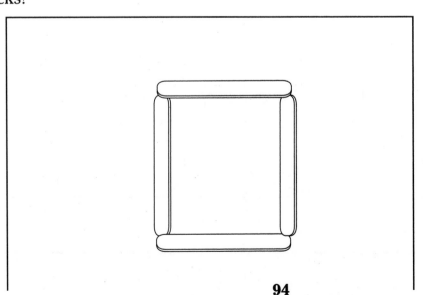

94

CHAPTER TWELVE

EGG CARTONS

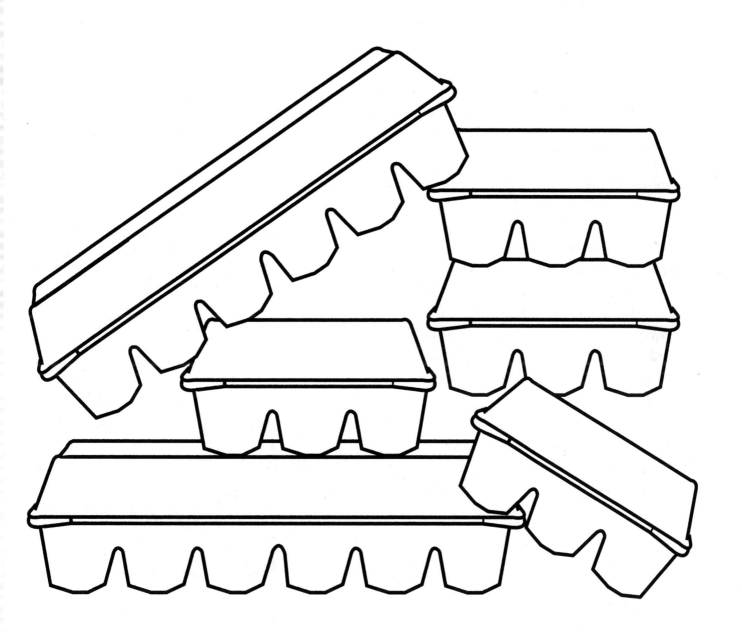

POPCORN COUNTING

KINDERGARTEN

Objective: To reinforce counting and numeral-recognition skills.

Materials: egg carton, popcorn kernels, glue, numeral cards (p. 97)

Construction:
1. Glue kernels of corn to the bottom of each egg carton section in the following number sequence:

3	5	6	2	4	7
2	7	4	3	6	5

2. Make a copy of the numeral cards, laminate, and cut out.

Directions for Student:
1. Count the number of popcorn kernels in each egg section.
2. Put the matching numeral card in each section.

NUMERAL CARDS

SCRAMBLED EGGS

FIRST GRADE

Objective: To reinforce number sequencing.

Materials: egg carton, number eggs patterns (p. 99), scissors

Construction:
1. Cut off the top of the egg carton and discard.
2. Duplicate the number eggs, laminate, and cut them out.

Directions for Student:
1. Look at the numbers on the egg-shaped cards.
2. Arrange the cards in the egg carton sections in numerical order, from the smallest number to the largest number.

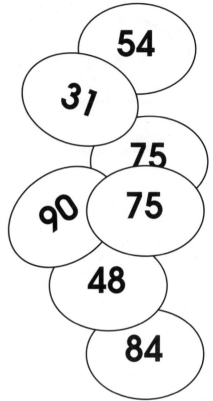

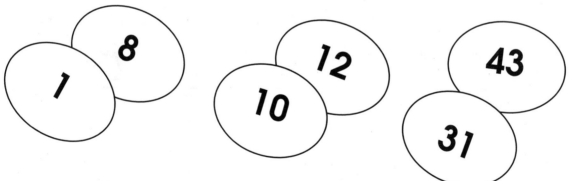

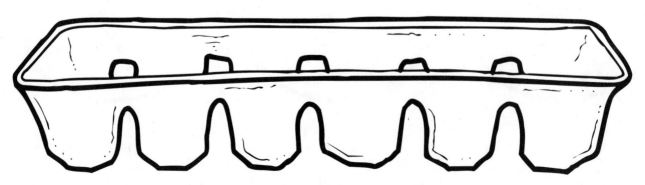

NUMBER EGGS PATTERNS

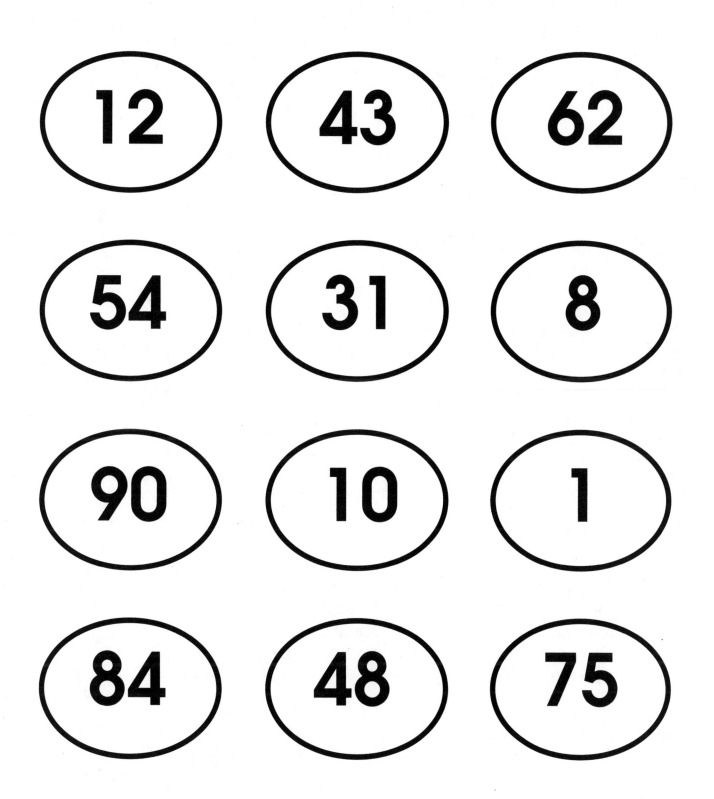

12   43   62

54   31   8

90   10   1

84   48   75

COUNTING EGGS BEFORE THEY HATCH!

SECOND GRADE

Objective: To reinforce understanding of ones, tens, and hundreds places.

Materials: egg carton, markers, egg patterns (p. 101), plastic bag, scissors, construction paper (blue, red, and green)

Construction:
1. Duplicate the egg patterns as follows: 12 blue eggs, 12 red eggs, and 12 green eggs.
2. Write these numbers in the egg carton sections:
   123, 456, 789, 987, 654, 321, 135, 680, 513, 896, 293, 310
3. On the blue eggs, write:
   100, 100, 200, 300, 300, 400, 500, 600, 600, 700, 800, 900
4. On the red cards, write:
   10, 10, 20, 20, 30, 50, 50, 80, 80, 80, 90, 90
5. On the green cards, write:
   0, 0, 1, 3, 3, 3, 4, 5, 6, 6, 7, 9
6. Laminate all of the eggs and cut them out. Store them in the plastic bag when not in use.

Directions for Student:
1. Look at the number in each egg carton section.
2. Find the eggs that show the correct number in ones, tens, and hundreds, and place the right three egg cards in each section. There will be one blue card, one red card, and one green card in each section, for example, for a section numbed 999 you would put in the 900 blue egg, the 90 red egg, and the 9 green egg.

EGG PATTERNS

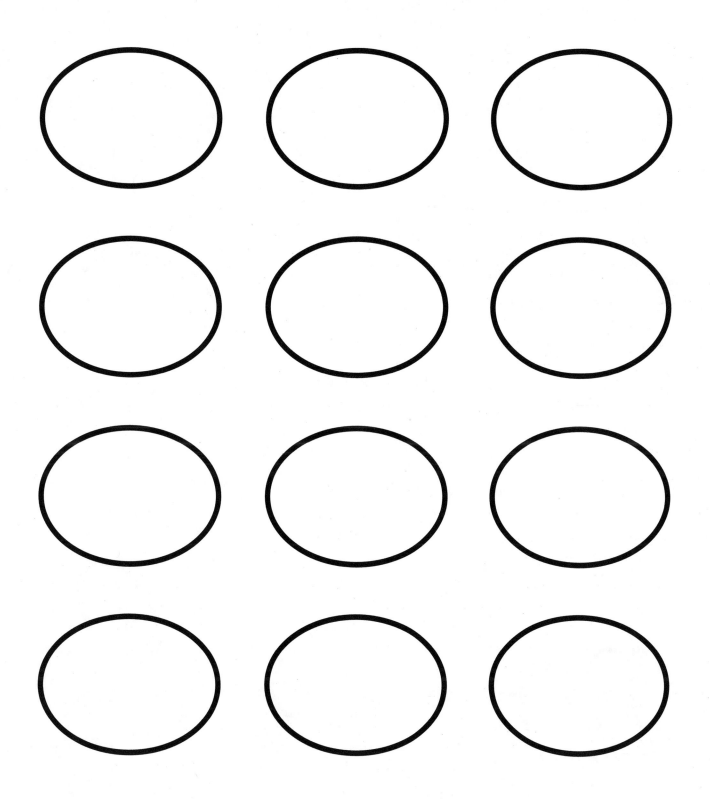

EGG-HEAD MATH

THIRD GRADE

Objective: To reinforce multiplication skills.

Materials: egg carton, marker, egg problem patterns (pp. 103-104), scissors, plastic bag

Construction:
1. Write each of these numbers in an egg carton section:
    0, 6, 12, 18, 24, 30, 36, 42, 48, 54
2. Duplicate the egg problem patterns, laminate, and cut out.
3. Store the egg shapes in a plastic bag.

Directions for Student:
1. Look at the problem on each egg.
2. Put each egg in the correct answer section of the carton.

Note: These problems concentrate on facts of 6. You can make up your own problems using the blank egg shapes on p. 101, to reinforce other multiplication facts.

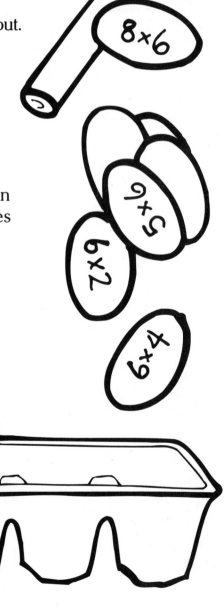

EGG PROBLEM PATTERNS

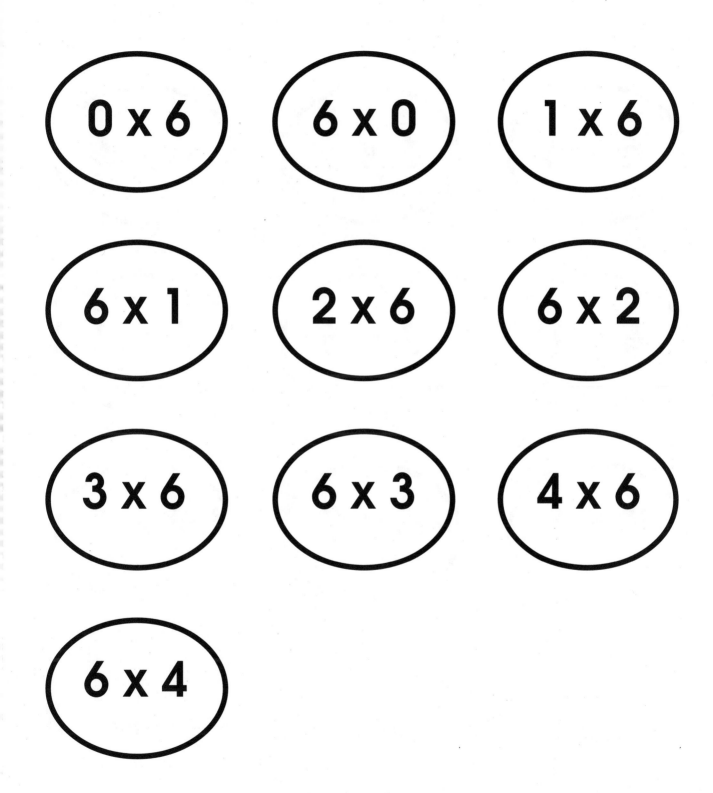

0 x 6   6 x 0   1 x 6

6 x 1   2 x 6   6 x 2

3 x 6   6 x 3   4 x 6

6 x 4

EGG PROBLEM PATTERNS

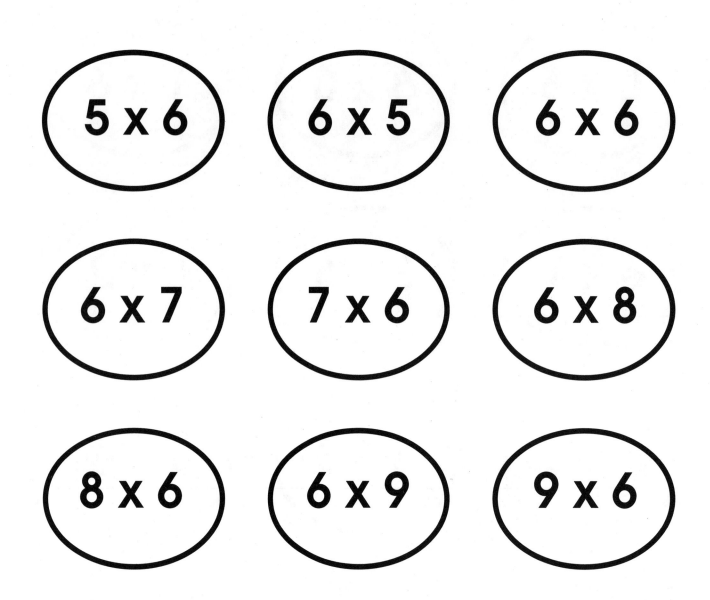

CHAPTER THIRTEEN

USED GREETING CARDS

PUZZLE PROBLEMS

KINDERGARTEN

Objective: To reinforce number recognition and provide counting practice.

Materials: 15 greeting cards (picture half only), number patterns (pp. 107-122), scissors, plastic bag, glue

Construction:
1. Duplicate the number patterns and cut out.
2. Glue one set of the patterns (number-number words or number-number patterns) onto the blank side of each greeting card.
3. Cut the cards in half (along the lines as shown) so that the numbers are on one side and the shapes or words representing the amounts are on the other.
4. Laminate the cards and store all pieces in a plastic bag.

Directions for Student:
1. Spread all of the pieces on a flat surface.
2. Match the numbers with the number patterns or number words.
3. Turn over the cards to check your work. If the picture halves go together, you are right.

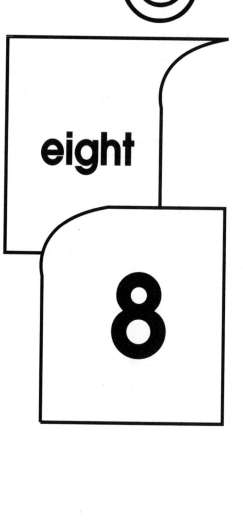

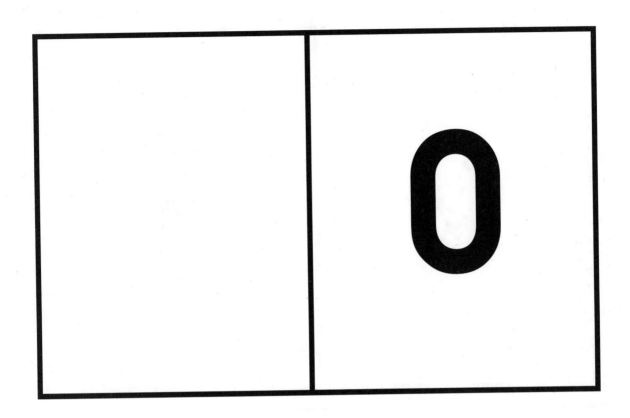

zero 0

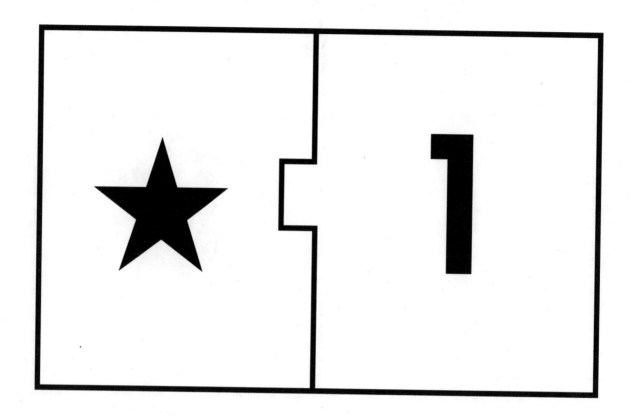

NUMBER PATTERNS

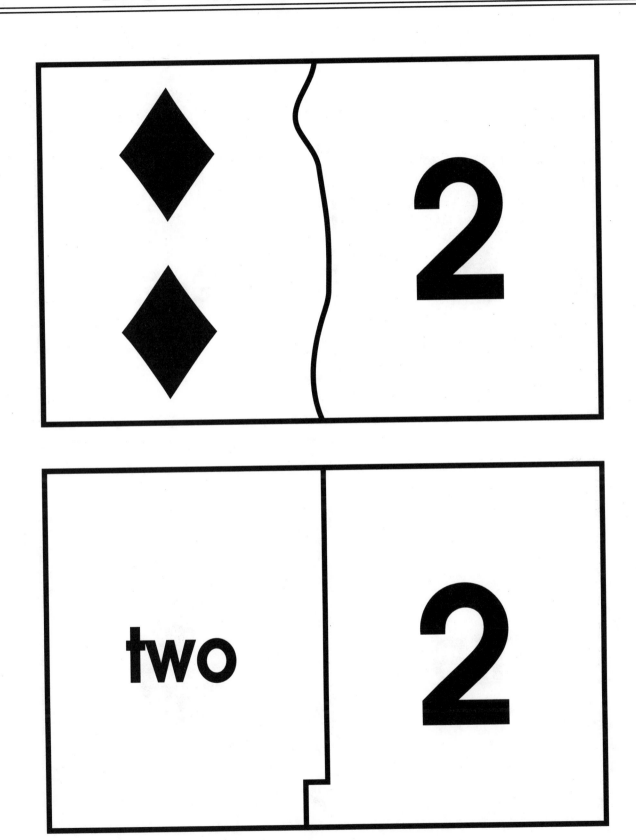

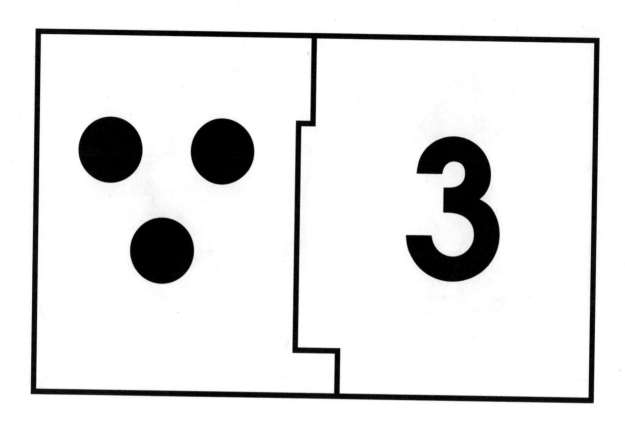

NUMBER PATTERNS

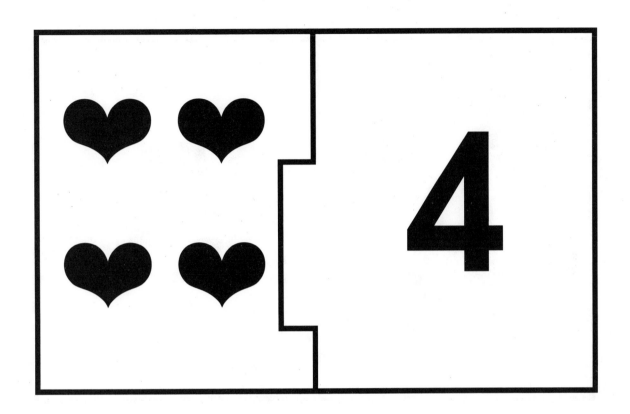

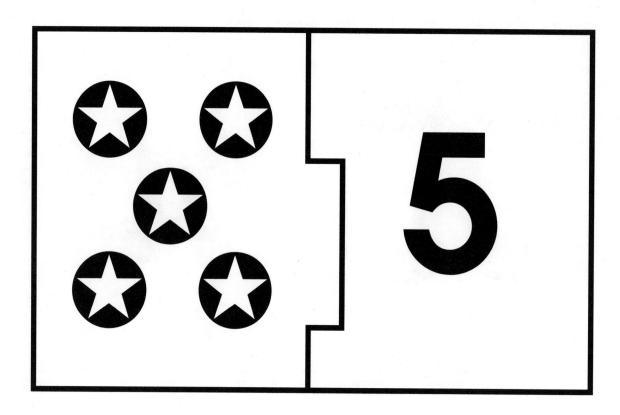

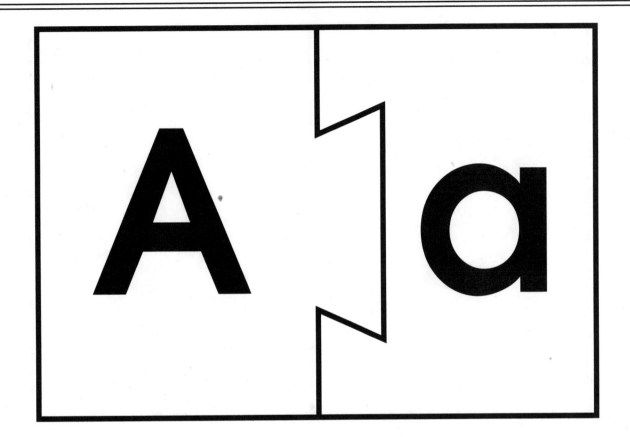

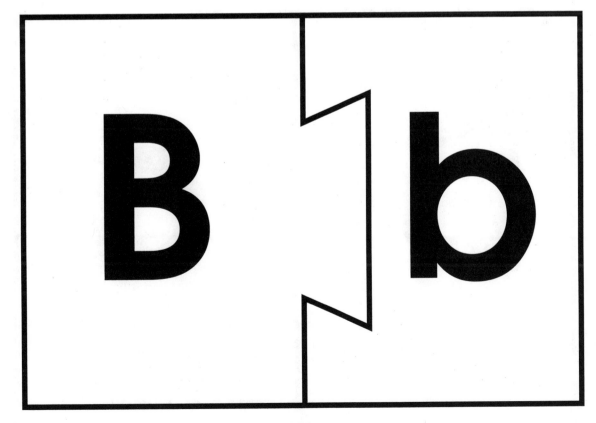

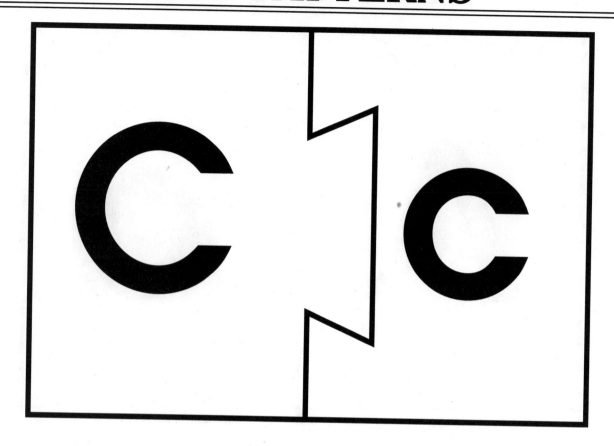

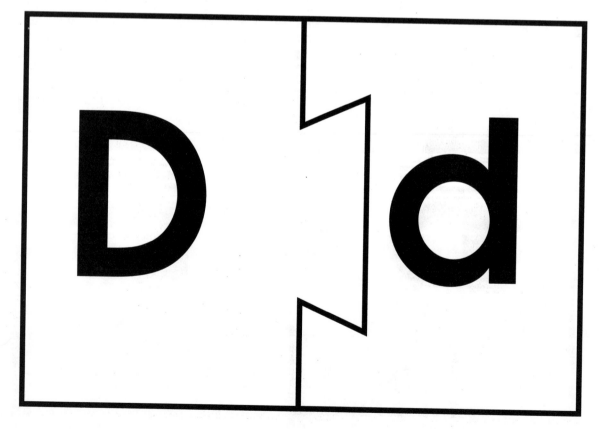

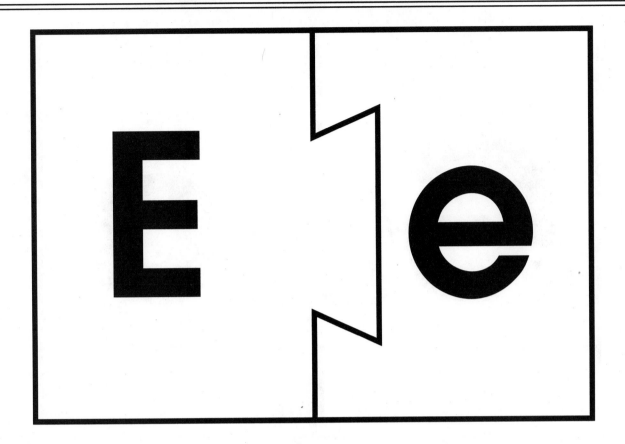

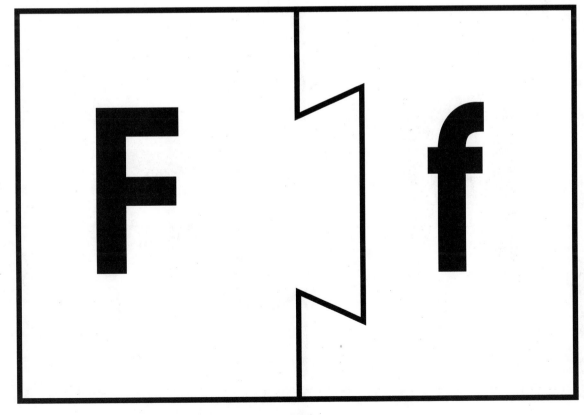

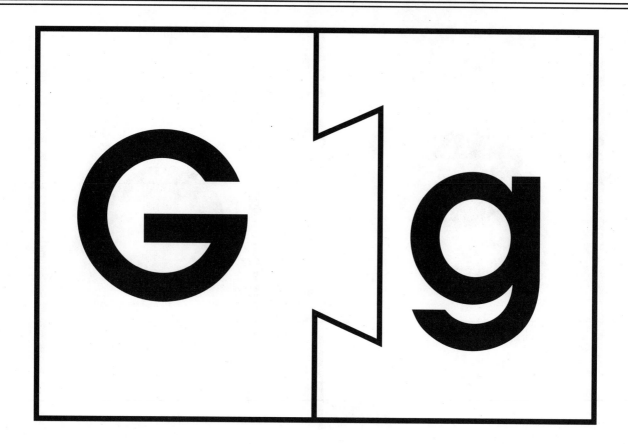

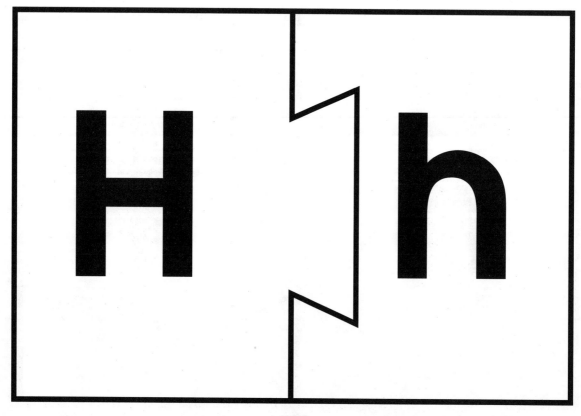

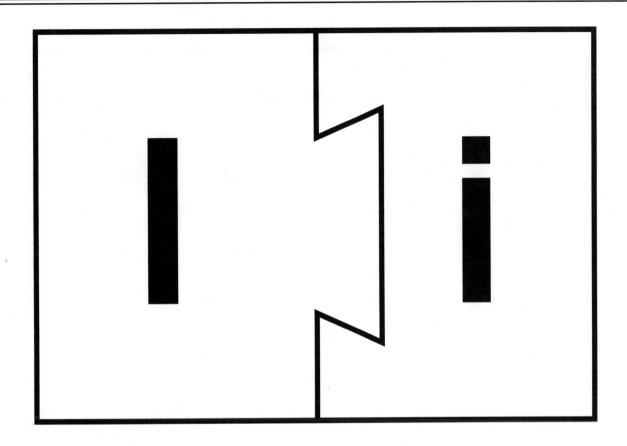

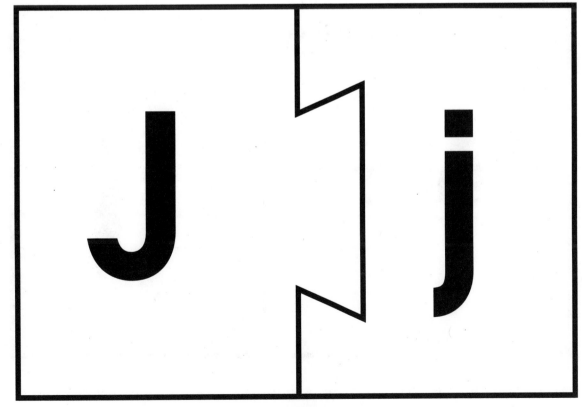

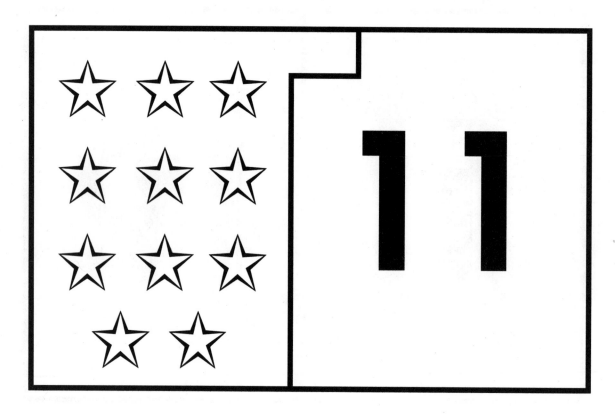

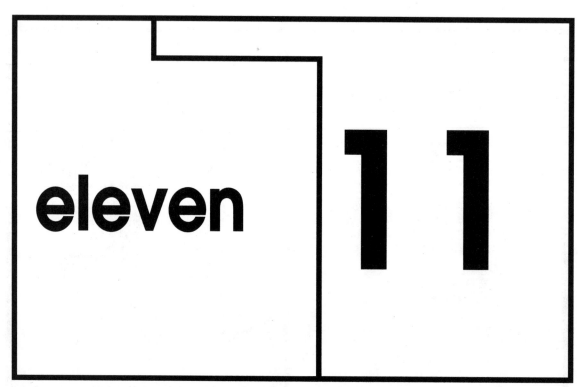

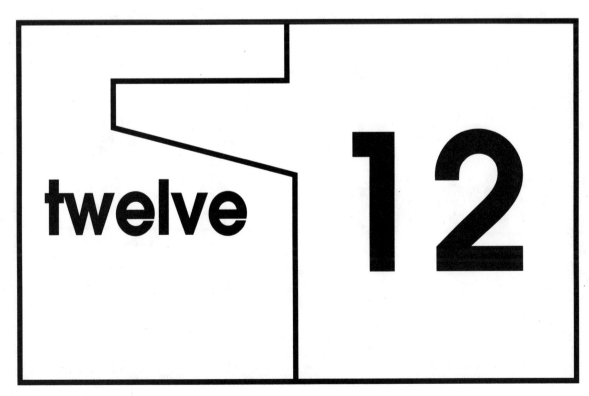

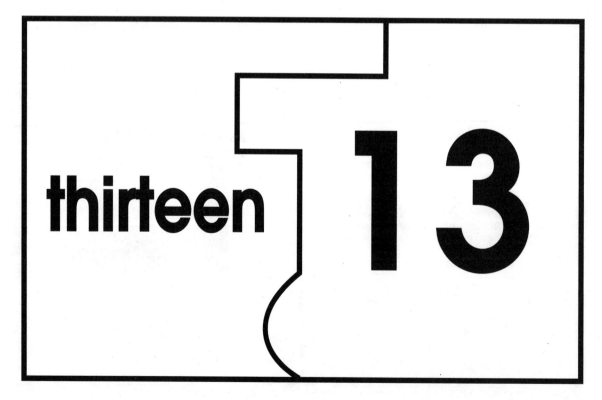

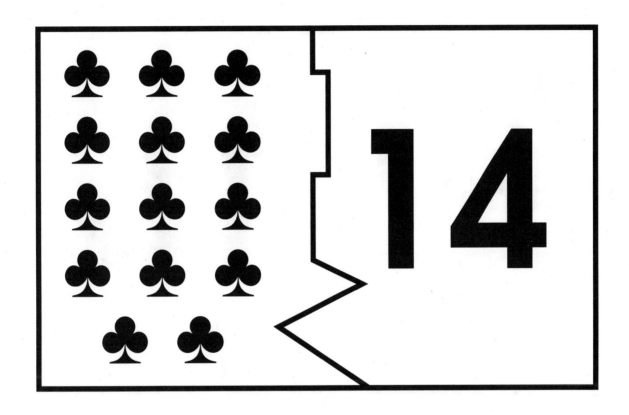

NUMBER PATTERNS

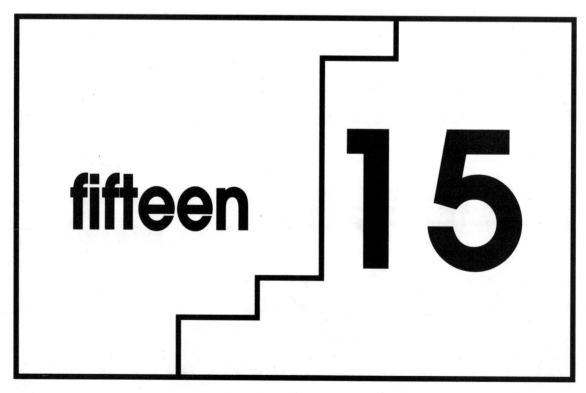

CLOCK CONCENTRATION

FIRST GRADE

Objective: To practice telling time.

Materials: 15 used greeting cards (picture half only), markers, clock patterns (pp. 124-125), glue, plastic bag, scissors

Construction:
1. Cut the cards in half. Also cut the clock patterns apart.
2. Glue a clock to one half of each card and the correct time to the other half.
3. Laminate all of the pieces, cut out, and store in a plastic bag.

Directions for Two Students:
1. Spread out all of the cards, with the clocks and times facing up.
2. Take turns trying to match each clock with its correct time card.
3. Turn over the cards to check yourself. If the pictures on the back go together, you are right.

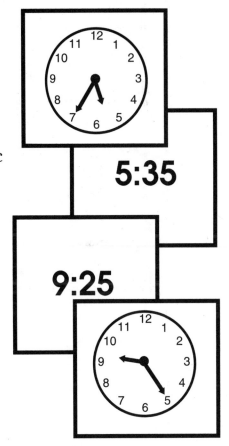

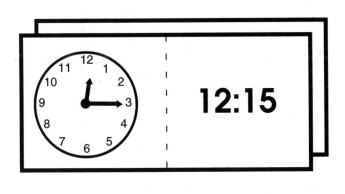

CLOCK PATTERNS

12:15

5:35

9:25

CLOCK PATTERNS

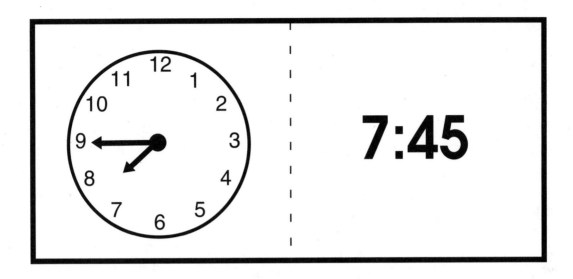

7:45

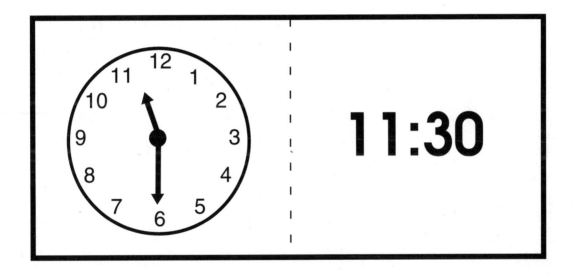

11:30

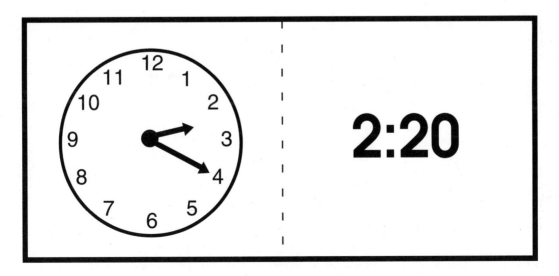

2:20

MONEY MATCH

SECOND GRADE

Objective: To reinforce addition skills.

Materials: 20 greeting cards (picture side only), markers, coin patterns (pp. 26 and 56), scissors, plastic bag, glue

Construction:
1. Cut the cards in half.
2. Write a money amount on one half, and glue pictures of coins to equal that money amount on the other half.
3. Laminate all cards and store in a plastic bag.

Directions for Two Students:
1. Spread out all the cards with the money amounts and coins face up.
2. Take turns matching the money amount cards with the correct coin cards.
3. Turn over the cards to check your work. If the pictures on the back go together, you are right.

$1.30

PICTURE PROBLEMS

THIRD GRADE

Objective: To practice division skills.

Materials: 15 greeting cards (all the same size; picture side only), marker, plastic bags, scissors

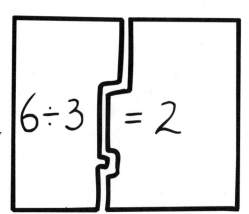

Construction:
1. Write a division fact on the blank side of each greeting card—write the problem on one side and the answer on the other side.
2. Cut the cards in half in a variety of ways: jagged, wiggly line, curved, etc.
3. Laminate all of the pieces.
4. Store the problems in one plastic bag and the answers in another.

Directions for Student:
1. Spread out the problem and answer cards.
2. Work out each problem and then find the answer.
3. Put the two pieces together and turn over. If the picture sides go together, you are correct. If not, try again.

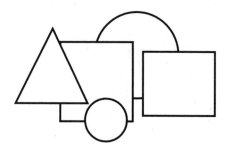

About the Authors

Brenda Morton

Brenda Morton is currently an elementary school principal, having previously taught kindergarten for 16 years. In addition to a degree in elementary education, she has a Master's Degree in Elementary Administration.

Bonnie Mertzlufft

Bonnie Mertzlufft is a full-time kindergarten teacher. She has a degree in Elementary Education with a special endorsement for kindergarten and a Master's Degree in Early Childhood Education. Bonnie has had a varied teaching career ranging from an at-risk group of 3-year-olds to a learning disabilities group with an age span of 5 to 18. She has had close to 20 years of teaching experience.

Virginia Woolf

Virginia Woolf teaches second grade at Bunker R-3 school, a position she has held for her entire teaching career. Virginia has a degree in Elementary Education from Mid-America Nazarene College in Kansas. She has taught for 21 years and has firsthand knowledge that hands-on activities work in her classroom with a range of students.